ETHIOPIAN CHURCH FORESTS

OPPORTUNITIES AND CHALLENGES FOR RESTORATION

AF880118

KERKBOSSEN IN ETHIOPIË

MOGELIJKHEDEN EN UITDAGINGEN VOOR RESTAURATIE

PROMOTER:

Prof. Dr. F.J.J.M. Bongers
Persoonlijk hoogleraar bij de leerstoelgroep Bosecologie en Bosbeheer, Wageningen Universiteit

CO-PROMOTEREN:

Dr. Demel Teketay Fanta
Regional Director, Forest Stewardship Council, African Regional Office (FSC Africa), Accra, Ghana

Dr. Ir. F.J. Sterck
Universitair Docent bij de leerstoelgroep, Wageningen Universiteit

PROMOTIECOMMISSIE:

Prof. Dr. M.J.A. Werger, Universiteit Utrecht
Prof. Dr. F. Berendse, Wageningen Universiteit
Prof. Dr. ir. B. Muys, K.U. Leuven, België
Prof. Dr. R. Boot, Tropenbos International, Wageningen

Dit onderzoek is uitgevoerd binnen de C.T. de Wit Graduate School Production Ecology & Resource Conservation

ETHIOPIAN CHURCH FORESTS
OPPORTUNITES AND CHALLENGES FOR RESTORATION

Alemayehu Wassie Eshete

PROEFSCHRIFT

Ter verkrijging van de graad van doctor
Op gezag van de rector magnificus
Van Wageningen Universiteit,
Prof. Dr. M.J. Kropff,
in het openbaar te verdedigen
op woensdag 31 oktober 2007
des ochtends te 11.00 uur in de Aula

Alemayehu Wassie Eshete (2007)
Ethiopian Church Forests: opportunities and challenges for restoration.

PhD thesis, Wageningen Univeristy, Wageningen, The Netherlands.
ISBN: 978-1-969702-11-2

Subject headings: species richness and diversity, structural composition, soil seed bank, seed predation, livestock grazing, microsite, management, regeneration, restoration, church forest, dry afromontane forest, Ethiopia

This study was carried out at the Forest Ecology and Forest Management Group, Centre for Ecosystem Studies, Wageningen University.

This Thesis is dedicated to my late parents Ato Sisay Wassie and W/o Munit Emiru whom I missed too early.

I would like also to dedicate to the Ethiopian Orthodox Tewahido Church and its scholars for generations of dedication and faithfulness to the church and its surrounding forests. Had not been to the conservation patronage of Ethiopian orthodox church, these forests would have gone long time ago.

CONTENTS

CHAPTER 1

GENERAL INTRODUCTION

General Introduction

Tropical dry forests are among the most exploited forest ecosystems of the world and have been increasingly fragmented (Cabin et al., 2002). The fragmentation results in small stands of forest bordered by open, often agricultural lands (Bustamante and Castor, 1998; Kollmann and Buschor, 2002; Fleury and Galetti, 2004; Saunders et al., 1991). Under the increasing pressure by the human population, via e.g. grazing activities or tree harvesting, natural regeneration may be hampered and, as a result, the persistence of the remnant forest patches and their indigenous species in many areas are threatened. This is particularly the case in the rapidly developing tropical areas such as in Ethiopia (Cotler and Ortega-Larrocea, 2006).

Dry Afromontane forests of Ethiopia have faced vast exploitation and almost all these forests have been converted to agricultural lands (EFAP, 1994; Demel, 1996; Tesfaye et al., 2003), except for small fragments that are left in the most inaccessible areas or around churches ("Church forests") (Bingelli et al., 2003; Alemayehu Wassie et al., 2005a; Aerts et al., 2006a; Bongers et al., 2006). The disappearance of the forests has been most drastic during the past 100 years. In the beginning of the 1900s, the forested area of the country was estimated at about 40% (EFAP, 1994) but now downscaled to 4.2 % of the land area (FAO, 2001). However, the small isolated patches of natural forest stands occurring around more than 30,000 churches still persist in a degraded landscape and might contribute to restoration, biodiversity conservation and provide many other economic and social benefits. The effectiveness of these church forests to provide "ecosystem services" for the landscape and serve as 'stepping stones' for restoration will depend on their long-term sustainability. Recent inventories show that populations of the tree species in the church forests are small, decreasing in extent over time and several tree populations appear to have no regeneration at all (Alemayehu Wassie, 2002; Bingelli et al., 2003; Alemayehu Wassie et al., 2005a). Population decline and failure of regeneration may contribute to a progressive ecological deterioration of church forests (Viana et al., 1997; Alemayehu Wassie, 2002; Bingelli et al., 2003). The overall objective of this PhD research is therefore, to assess forest community structure and composition of the church forests, investigate major problems for regeneration of woody species and thus explore the possible options for conservation and forest restoration in Northern Ethiopia.

Ethiopian churches and their enveloping remnant forests

In the northern highlands of Ethiopia, patchy remnants of old-aged Afromontane forests can be found mainly around the Ethiopian Orthodox Tewahido Churches (EOTC). Forests in other areas have been completely destroyed and converted into farms and grazing lands over centuries. Hence, when a traveler sees a patch of indigenous old-aged trees in the northern highlands of Ethiopia, he/she can be sure that there is an Orthodox Church in the middle. They are visible from a great distance, with a majestic appearance, usually built on small hills "overlooking" the surrounding villages. The local people call these churches with the surrounding trees as "debr" or "geddam". "Debr" or "Geddam" is seen by the followers as the most holy place religiously as well as a respected and powerful institution socially.

EOTC, an indigenous and integral Christian Church of Africa, is one of the oldest Churches in the world (Aymro and Motovu 1970; Alemayehu, 2002). The word 'Tewahido' is the Ethiopian term meaning 'made one'. The church considers this word the best expression conveying the faith of the church, as it emphasizes the inseparable unity of the Godhead and Manhood in the Person of Christ. The EOTC is numerically the largest of the five non-Chalcedonian Eastern Churches – the Coptic, the Ethiopian, the Syrian, the Indian and the Armenian, which were historically called "The Oriental Orthodox Churches". It has over 40 million followers, 500,000 clergies and 35,000 churches in Ethiopia. In addition to its religious activities, EOTC has also a long history of conservation of forest resources, which usually envelop the churches. Although the main purpose of churches is as places for worship, burials and meditating religious festivals, they also provide valuable, often unique, and secured habitats for plants and animals, and green spaces for people.

The Church perceives nature in a holistic manner. In their concept, nature includes among others, human beings, forests/trees, animals, water, other land features and the nation as a whole. Pilot studies suggest that church forests might be relics or blueprints of ancient and largely lost forest ecosystems, that they are hotspots of biodiversity for indigenous species, and therefore they might serve as priority areas for in-situ conservation and meanwhile maintain as prestigious religious sites (Alemayehu, 2002; Bingelli et al., 2003; Alemayehu Wassie et al., 2005a&b; Bongers, et al., 2006). The local communities do have higher respect and thrust in the EOTC than other local institutions, which has made the church the central institution and platform for socio-economic issues of the people. The positive attitude to the resources protected by the church and the acceptance of the church tradition could thus be an opportunity for forest ecosystem conservation and restoration (Alemayehu Wassie, 2002).

From vegetation sampling done on eight of these churches in South Gondar (Northern Ethiopia), it was found that forests that surround the church have an area ranging from 1.6 ha to 100 ha. The total number of woody species and families in each of the eight churches ranged from 18 and 22 to 22 and 42 respectively. The plant density ranged from 731/ha to 2250/ha and basal area (calculated from woody plants with dbh $\geq$ 10 cm) ranged from 25 m^2/ha to 110 m^2/ha. The South Gondar indigenous woody species profile was compared as a checklist with the species recorded in the sampled churches. Accordingly from the total 125 woody species found in the zone, 81 species were recorded in the eight sampled churches and 16 species were recorded which had not been listed in the checklist (Alemayehu Wassie, 2002; Alemayehu Wassie et al., 2005a). In another study, 38 sites were investigated and the forests ranged between 0.5 and 50 hectares and on average contained 41 species of woody plants (range: 21-68) (Bingelli et al., 2003). This suggests that EOTC forests might strongly contribute to conserve woody species diversity in Ethiopia and in Northern Ethiopia in particular.

Habitat fragmentation and restoration

Habitat fragmentation is often defined as a process during which “a large expanse of habitat is transformed into a number of smaller patches of smaller total area, isolated from each other by a matrix of habitats unlike the original” (Wilcove et al., 1986). Four effects of the process of fragmentation on habitat pattern are distinguished: (a) reduction in habitat amount, (b) increase in number of habitat patches, (c) decrease in sizes of habitat patches, and (d) increase in isolation of patches (Fahrig, 2003). Fragmentation invokes more than habitat removal because it causes not only loss on the amount of habitat but also changes in the properties of the remaining habitat (van den Berg et al., 2001). Its negative effects are likely due to two main causes: First, fragmentation results in too small patches to sustain local populations of plants at some point. Second, fragmentation also creates more edge for a given amount of habitat, increasing the probability of individuals leaving the habitat and entering the matrix where overall mortality rate is very high (Fahrig 2003). As a result fragmentation has been described as one of threats to global biological diversity (Hill and Curran, 2005).

Ecological restoration is an intentional activity that initiates or accelerates the recovery of an ecosystem with respect to its health, integrity and sustainability.

Frequently, the ecosystem that requires restoration has been degraded, fragmented, transformed or entirely destroyed as the direct or indirect result of human activities (SER, 2002; Falk et al., 2006; Van Andel and Aronson, 2006). In some cases, these impacts to ecosystems have been caused or aggravated by natural agents (Miller et al., 1995; Eshetu Yierdaw, 2002; SER, 2002). An apparent dichotomy is often erected between conservation and restoration indicating that they are considered to be alternative options. However, the dichotomy doesn't exist, since restoration activities should be placed with in the broader context of sustainable land use and conservation (Hobbs and Harris, 2001).

There are three concepts used in restoration ecology based on alternative goals. These alternatives are similar in that they all aim to re-create productive and functional ecosystems on an otherwise degraded landscape, but they differ in the extent to which they seek to reestablish the original ecosystem (Lamb et al., 1997; Harris and Van Diggelen, 2006). (i) *Reclamation*: actions that usually involve site amelioration that permits vegetation to establish and colonize the site and increase its utility or economic value. It rarely refers to indigenous ecosystems as a model and, rather, uses only exotic species to overcome degradation. (ii) *Rehabilitation*: actions that seek to repair damaged ecosystem functions, particularly productivity, quickly. Indigenous species and ecosystem structure and function are the targets for rehabilitation but also exotic species might be included. (iii) *Restoration*: actions that attempt to restore a degraded ecosystem to its presumed original condition. There are two scopes in restoration; *restoration sensu stricto* and *restoration sensu lato*. Restoration *sensu stricto* refers to actions that lead to the full recovery of an ecosystem to its pre-disturbance structure and function, while restoration *sensu lato* refers to actions that seek to reverse degradation and to direct the trajectory in the general direction of one aspect of an ecosystem that previously existed on the site (Lamb et al., 1997; Walker and Moral, 2003). In this study, it is focused on the third concept, i.e. "restoration", but in sensu lato sense.

Restoration in this sense, i.e. towards the original ecosystem including the indigenous biodiversity, is hardly practiced in Ethiopia, where most emphasis has been on reclamation and rehabilitation efforts (Bendz, 1986; Tefera Mengistu, 2001; Emiru Birhane, 2002). Restoration attempts in reference to the tremendously scattered forest remnants around churches are still lacking in Northern Ethiopia. More specifically, this thesis will focus on the conservation of the current church forests and, at the same time, restoration of these forests in the surrounding degraded landscape.

Restoration attempts to return an ecosystem to its historic trajectory. An ecological trajectory is one that describes the developmental pathway of an ecosystem through time. The general direction and boundaries of that trajectory can be established through studies on comparable intact ecosystems and remnants, information about regional environmental conditions, and analysis of other ecological, cultural and historical reference information (SER, 2002). In some cases, even though it seems that all the original biological community has been destroyed, small areas containing most of the indigenous species remain scattered to make restoration possible (Cunningham and Saigo, 1995). Even if forest remnants are small they may have important biological, economic, and social values (Laurance and Bierregaard, 1997). These remnants are, in essence, the raw materials available to rebuild the ecosystem (MacMahon, 1997; Eshetu Yierdaw, 2002), and here we study whether the church forest can acts as such in the degraded Northern Ethiopian landscape.

Regeneration: a key ecological process for conservation and restoration of remnant forests

The status of ecological functions of a forest ecosystem can be investigated by selecting an ecological process vital to forest function that deemed to be potentially vulnerable to fragmentation (Crome, 1994). In this respect, the regeneration of trees can be amongst such processes (Grime and Hillier, 2000). If native long-lived trees are unable to survive and regenerate in a given forests, then there is little hope for maintaining any semblance of normal forest functioning in the long term (Harriington et al., 1997; Janzen, 1986). Therefore, the effectiveness to maintain (conservation) and expand (restoration) church forests will depend mainly on the success of natural regeneration of the component tree species.

Earlier pilot studies have already shown that natural regeneration is not successful for many indigenous species (Alemayehu Wassie, 2002; Bingelli et al., 2003; Alemayehu eassie et al., 2005a). These studies showed that most of the tree species had small population sizes and with no or very few individuals in the lowest diameter classes as revealed from their diameter class distributions (Alemayehu Wassie, 2002). Tree populations in church forests are not only small but also decreasing in size in several cases. Some species had estimated population sizes of fewer than twenty individuals, which may lead to genetic drift (Nason et al., 1997; Alemayehu Wassie, 2002; Bingelli et al., 2003). Several tree populations appear to have no regeneration, and the remaining individuals may be reproductively non-functional. All these may contribute to a progressive ecological deterioration of church forests (Viana et al., 1997; Alemayehu Wassie, 2002; Bingelli et al., 2003). A wide variety of factors might contribute to the further degeneration of church forests, and yet it is not known how these factors contribute to the failure of regeneration. Moreover, the major challenge will be to control those factors such that regeneration becomes successful, inside and outside the forest. This major challenge receives the major focus in the presented studies.

Factors affecting regeneration of remnant forests

Factors that potentially influence regeneration at the early stage are those that determine the probability of seedling establishment and those that affect seedling survival and growth (Fig 1.).

Seedling establishment: refers to the ecological processes of seed germination and seedling establishment (emergence). Seedling establishment is mainly determined by seed availability and microsite conditions where the seedlings are to be established.

Seed availability: There could be three possible sources of seeds for germination: local seed rain, soil seed bank and dispersal from outside. There may not be sufficient input from seed rain both in quality and quantity in church forests. Reduced seed rain may be attributed to fragmentation due to changes in tree phenology as a result of increased mortality of reproductive individuals and/or genetic drift that might lead to a reduction of flowering (or setting of fruits and seeds); fewer animals for pollination and seed dispersal; and increased pre-dispersal seed predation. The availability of seeds from the soil seed bank depends on the ability of species to maintain persistent soil seed bank. Furthermore seed availability can be aggravated by intense post-dispersal seed predation. Seeds represent a particularly concentrated source of potential food of high nutritive quality to any organism able to exploit it. Therefore, it is not surprising to find that in many plant species, a large proportion of seed production is lost to predation (Fenner and Kitajima, 1999). Predation reduces the chances for reproduction. Especially, when regeneration is limited by seed numbers, the consequence of seed predation on plant regeneration and population would be very significant (Crawley, 1992; Hau, 1997; Fenner and Kitajima, 1999). It may also influence the genetic makeup of the plant population by differential selection of seeds even when it has little effect on recruitment (Benkman, 1995; Fenner and Kitajima, 1999). Therefore it is hypothesized that due to reduced seed rain and lack of persistent soil seed bank aggravated by intense seed predation, limited seed availability might be a bottleneck for seedling establishment in church forest (Fig 1.).

Unfavorable Microsite: In some instances the available seeds may not able to establish seedlings due to unfavorable microsite/environment for seed germination. Thick forest litter layer decreases germination and seedling emergence through shading, biochemical effects and physical obstruction to the emergence of a seed's cotyledons and radicle. Physical obstruction may either prevent seedling emergence or force seedlings to allocate more stored energy to hypocotyl growth in order to penetrate the litter layer, leaving less energy for allocation to the radicle and cotyledons. Seeds may therefore germinate but exhaust their energy reserves before emergence from the soil and litter layer. In contrast, thick litter cover protects large seeds from desiccation damage and predation. Although litter cover may have both negative and positive effects, a meta analysis of 32 published studies found that the overall effects of litter are negative on both seed germination and seedling establishment (Kitajima and Fenner, 2000; Ellsworth, et al., 2004). Therefore thick litter layer could be one of the problems that hamper seedling establishment in church forests. Compacted top soil due to repeated trampling by animals could also hamper germination (Smit et al., 2006a&b). Drastic variability and fluctuations in the microenvironment such as relative humidity, temperature and soil moisture and unfulfilled optimal requirements like light and moisture could also hamper germination and successful establishment. An unfavorable microenvironment may lead to either seed mortality or seed dormancy and deposition into the soil (as part of the soil seed bank) depending on the type of species (Bazzaz and Pickett, 1980; Vazquez-Yanes and Orozco-Segovia, 1993; Demel Teketay and Granström, 1997). In general it is hypothesized that due to unfavorable microsite conditions for successful germination, seedling establishment is hampered in church forests (Fig 1).

Seedling survival and growth: once seeds have germinated and seedlings are established the next challenge is to survive, grow and join sapling and adult stage. Seedling mortality may increase due to unfulfilled optimal requirements like light and moisture. Particularly in fragmented forests a consequence of drastic variability and fluctuations in the physical microenvironment (relative humidity, wind and temperature) is crucial. Physical damages and destruction due to trampling and grazing, competition by weeds and herbs, which deprives the young seedlings of light and soil resources and raises seedling mortality (Kitajima and Fenner, 2000), and diseases and pests could be another factor increasing mortality. Therefore, it is expected that due to unfavorable microsite, higher risk of physical damage and sever competition of weeds, seedling survival and growth has become lower in church forests (Fig 1.).

Each of the above underlying processes, which are responsible for failure of seedling establishment and increased seedling mortality, are predisposed by external factors. Among the possible factors that may affect these interplay ecological processes and in turn regeneration could be: rainfall gradient/altered hydrological regimes, soil type/nutrient runoff, altitude, size of the forest, gaps/canopy cover, anthropogenic disturbances (livestock grazing and tree cutting) and edge effects (Benitez-Malvido, 2001). In this study, we focus on how forest size, edge effects, canopy openness, and livestock grazing influence seedling establishment and seedling survival and growth (Fig 1.)

Forest size: In most cases, forest fragmentation results in isolated and small forest patches. It has considerable potential for rapidly reducing population sizes of some or all of the woody species in the communities and increasing isolation of their population and animal vectors, such as pollinators and seed dispersers (Powell and Powell, 1987; Rylands and Keuroghlian, 1988; Benitez-Malvido, 2001; Young et al., 1996; Lepsch-Cunha et al., 2001). This reduction may, in turn, lead to demographic constraints, impairment of positive interactions or alleviation of pressure from negative interactions. The smaller the population, the higher the chance that individuals will mate with close relatives, which may produce few offspring and/or weak or sterile offsprings, which is known as inbreeding depression or endogamy. Out-breeding depression is the opposite trend: the paucity of mates may lead to hybridization with nearly genetically incompatible individuals (say, different species from the same genus, which often produces vigorous but sterile hybrids) (Desouza et al., 2001). Small populations may also have their fitness severely affected by genetic drift (Nei et al., 1975; Desouza et al., 2001). As a result, the size of the forest can influence the quality and quantity of seed rain and hence seedling recruitment. Benitez-Malvido (1998) documented lower seedlings densities in small (<1 ha) relative to large (>10 ha) tropical forest fragments. Therefore, the sizes of church forests could have an impact on the regeneration success and diversity of woody species and in more particular we expect that the smaller the size of the forest, the lower is the seed availability via reduced seed rain, and the lower is the chance to get favorable microsite for seedlings growth and survival which ultimately reduces seedling density (Fig 1.).

Edge effects: One of the most obvious features of fragmented landscapes is a drastic increase in the amount of forest edge. The margins of forest fragments/church forests are usually abrupt transition from forest to pastures, crops, or other modified habitats. As the size of the forest is diminished, the proportion of edge exposed to other habitats and its effect increases. Edge effects can be loosely classified into physical and biotic effects (Laurance and Bierregaard, 1997). Physical edge effects includes elevated wind turbulence and temperature variability, lateral light penetration, and reduced humidity, all of which result from the close proximity of a harsh external climate in the surrounding matrix. The temperature, moisture and humidity may fluctuate drastically and suddenly within a short time period, which results in an unstable and unfavorable microclimate. Biotic effects can be extraordinarily diverse, and include the proliferation of secondary vegetation along forest margins, invasion of weedy or generalist plants and animals, alteration of ecological processes such as nutrient cycling and energy flows, and a myriad of other ecological changes. Some studies showed that relative humidity and soil moisture varied along transects running from the edge to the forest interior (e.g. Kapos, et al., 1997). Kapos (1989) reported edge effects, including elevated air temperature and increased vapor pressure deficit, extending at least 60 m into a 100 ha forest fragment. Another study showed that microclimate edge effects penetrate about 30 m into a small (20 ha) forest remnant under both dry and wet conditions and such trends had major implications for seed germination and seedling establishment (Turton and Freiburger, 1997). Seedling density increased significantly with distance from the edge within 100 and 10 ha fragments, with higher increases for the 100 ha fragments (Benitez-Malvido, 1998). Therefore, edge effects can have major impacts on the ecology of church forests and, thus, could affect the regeneration of most of the woody species and, ultimately, change of populations of woody species in church forests. We hypothesized that at the edge of the forest success in seedling establishment and seedling survival and growth is reduced (Fig 1.).

Canopy openness: Humans activity might affect the forest structure by cutting trees and creating gaps in the canopy. Gaps are temporary openings or plant-free spaces in vegetation, which provide an area with lower resource competition of some form for a period of time. Canopy gap is defined as 'a hole in the forest extending through all levels down to ... 2 m above the ground' (Brokaw, 1982) and canopy gaps vary in size from no (or very small) gap to very large gaps. Obviously, canopy gaps are always found to have higher light intensities at the soil surface than closed vegetation in forests (Chazdon & Fetcher 1984; Bullock, 2000). The higher light intensities may however reduce the water resources by decreasing surface soil moisture while water availability in deeper soil layer may be higher due to less vegetation that transpires (Denslow, 1980; Collins et al., 1985; Vitousek and

Denslow, 1986; Bradshaw and Goldberg, 1989) and may also lead to increase daytime soil temperature and lower nighttime temperatures (Thompson et al., 1977; Denslow, 1980; Vitousek and Denslow, 1986). Gap creation by plant death often results in an increase in dead biomass within the gap. Decomposition and mineralization of this biomass sometimes increase soil nutrient concentrations (Denslow, 1980; Bullock, 2000). The enhanced resource availability in gaps has a paramount effect on the regeneration of tree species. However, whether it has negative or positive effect may depend on the most critical limiting resource, in particular where light or water is limiting regeneration (Bullock, 2000). We hypothesized that where light is a limiting factor, seedling establishment and seedling survival and growth might improve in gaps but the reverse pattern might be observed where moisture is a limiting factor (Fig 1.). However the response of the species might differ based on specific light and moisture requirements.

Livestock grazing: Anthropogenic disturbances may contribute to regulate the regeneration dynamics, structure and floristic composition of tropical woodlands and montane forests. It has been suggested that a light level of grazing intensity can increase tree regeneration by removing competitive vegetation, removing fire hazard and fertilizing by droppings. Large herbivores can also create patches of bare ground through grazing and trampling. These disturbances, however, may create "safe sites" for seeds to germinate (sensu Harper, 1977; McEvoy et al., 2006). On the other hand, frequent but low-intensity disturbance, mainly livestock grazing and browsing, may involve the combined effect of multiple factors and strongly affect forest structure and the ability of understorey species to regenerate (Neptalí et al., 2001). Natural regeneration could be scarce, presumably, due to livestock grazing that results in high fruit/seed and seedling predation. Unrestricted animal movement may also destroy seeds and press them deep into the soil from where they are not able to emerge (Mwendera and Mohamed Saleem, 1997; Hulme and Borelli, 1999; Smit et al., 2006a, b). Trampling and grazing of seedlings may lead to low seedling survival (Carolina and Javier, 2001). In some areas, it was observed that the local people allow their cattle in the church forests to rest under the shade of the trees to protect them from strong sunrays and storms (Alemayehu Wassie, 2002). Livestock-induced disturbances might be among the major factors constraining regeneration and recruitment of woody species and contributing, ultimately, to the decline of woody species populations in church forests. In this study, we hypothesized that the destructive effect of livestock grazing outweighs its positive effect and thus hampers seedling establishment and seedling survival and growth in church forests (Fig 1.).

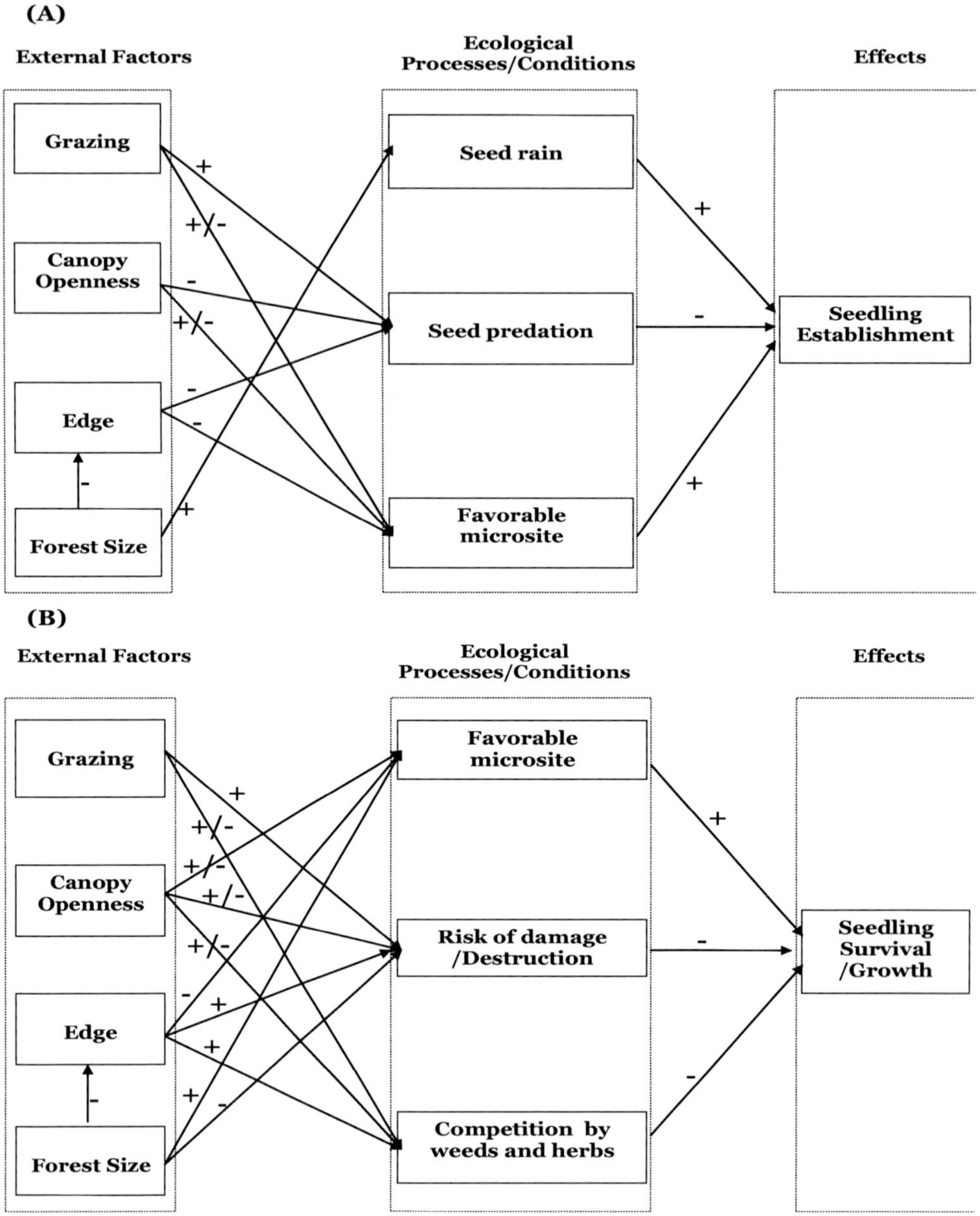

Figure1. Schematic representation of interaction of external factors versus ecological processes affecting (A) Seedling establishment (B) Seedling survival/growth of woody species in Ethiopian church forests. Expected positive effects/relations are indicated by a + sign along the arrows, and negative effects/relations by a – sign.

Restoration interventions

A common goal for the restoration of any natural ecosystem is to recover autogenic processes to the point where assistance from restorationists is no longer needed (SER, 2002; Van Andel and Aronson, 2006; Falk et al., 2006). Therefore, the major challenge is to facilitate natural regeneration of or to reproduce the existing forests / species and create habitats where they can thrive and re-create the forest. In fact, according to Janzen (1988) habitat restoration is primarily the initiation, growth and coalescence of remnant habitat fragments.

Four possible methods for restoring natural vegetation are (i) natural regeneration, (ii) direct seeding, (iii) planting seedlings, and (iv) incorporation of restoration goals in plantation programs (Lamb et al., 1997; Vallejo et al., 2006). The mechanism of natural regeneration is based on vegetative recovery of buried seeds and chance of seeds immigration from the outside. It is very slow and the outcome is unpredictable. Artificial seed and seedling introduction, and exposure of the buried seeds, on the other hand, might more rapidly facilitate the restoration process. Often, re-introducing a keystone species (one that is critical in determining the structure or function of an ecosystem) will help to restore a range of organisms with which that species once coexisted. Where restoration is being attempted by introducing seeds and seedlings, the key stone species should attract seed disperses, which in turn bring other plant species into the area (Lamb et al., 1997; Van Andel and Grootjans, 2006). This may also accommodate establishment of *in-situ* and *ex-situ* conservation places and seedling nursery for the retrieval of germplasm and safeguard the continuation of generation of these species. Moreover, some studies indicated that plantations may enhance the recruitment, establishment and succession of native woody species by functioning as foster ecosystems (Lugo, 1997; Eshetu Yirdaw, 2001; Feyera Sebeta et al., 2002; Mulugeta Lemenih and Demel Teketay, 2005). The use of plantations to restore diverse tree vegetation seems paradoxical, but it works because the manager can match species to particular site conditions and thus overcome limiting factors that prevent regeneration of species- rich forests on degraded sites. Once a forest is established, micro-site conditions change and wildlife is attracted. Animals are likely to disperse tree species from surrounding forest patches and regeneration of shade-intolerant species can be inhibited. Therefore, though the aim of these plantations at the beginning could be production oriented, ultimately they can end up taken over by the native communities (Lamb et al., 1997).

Objectives and research questions

Church forests are expected to be the last refuges of many species and may act as stepping stones for restoration. However, information on these forests is scarce. This PhD research assesses forest community structure and composition of the church forests, investigates major bottlenecks for regeneration of woody species, and explores opportunities and challenges for restoration. In this thesis, the following major questions are addressed:

1. How do forest structure, species composition and biodiversity vary across church forests, and what are the major factors driving such variations? Factors considered are altitude, forest size, livestock grazing intensity and wood harvest intensity. This issue is studied by comparing 28 church forests, distributed over the study area.

2. What are the major bottlenecks in the regeneration of woody plants in church forests? These bottlenecks are studied for one to seven church forests and a major focus is given to the effects of soil seed bank (chapter 3), post-dispersal seed predation (chapter 4) on seedling establishment, and of livestock grazing (chapter 5), microsite gradients and management interventions (chapter 6) on seedling establishment, seedling survival and growth.

Study area

The study was conducted in South Gondar Administrative Zone (SGAZ), one of the Administrative Zones in the Amhara National Regional State, Northern Ethiopia. Its geographical location is between 11° 02′ - 12° 33′ N and 37° 25′ - 38° 41′ E with an altitude range of 1500-4231 m (Fig. 2) . The capital town, DebreTabor is located 600 km North of Addis Ababa. SGAZ has an area of about 14,299 km^2 of which about 9% is classified under 'Kolla' (500 –1500 masl), 73% under 'Woynadega' (1500-2300 masl), 16% under 'Dega' (2300- 3200 masl) and 2% under 'Wurch' (> 3200 masl) Agro-Climatic Zones. SGAZ has 10 Administrative Districts, known locally as 'Woredas' with a total population of 2,426,123 (CSA, 2005) of which 1,231,219 were males and 1,194,904 were females. Estimated population density is 169.21 people per square kilometer. The average family size is about five while the average rural household has 1 hectare of land (compared to the national average of 1.01 hectare of land and an average of 0.75 for the Amhara Region) and the equivalent of 0.6 heads of livestock (CSA, 2005). In SGAZ, there are 1404 churches, and 95% of total population is member of the EOTC (SGAZDPED, unpublished).

The rainfall is characterized by a bimodal distribution with the major rainy season being from June-August ('Kermit') and a lesser amount of rainfall occurs from March-May ('Belg'). The annual average rainfall varies between 700 – 1300 mm and the annual average temperature ranges between 9.3 - 23.7 °C. The topography comprises uneven and ragged mountainous highlands, extensive plains and also deep gorges. It is one of the degraded and eroded areas in the Amhara region. The common soil types are Vertisols, Cambisols, Rigosols, Leptosols, Fluvisols and Arenosols. According to the Zonal Department of Agriculture (SGAZDA unpublished) the total forest cover in SGAZ is 20,882 ha, comprising 16,660 ha of natural forest and 4222 ha of man-made plantations. The forest cover accounts for only 1.4% of the total area of SGAZ.

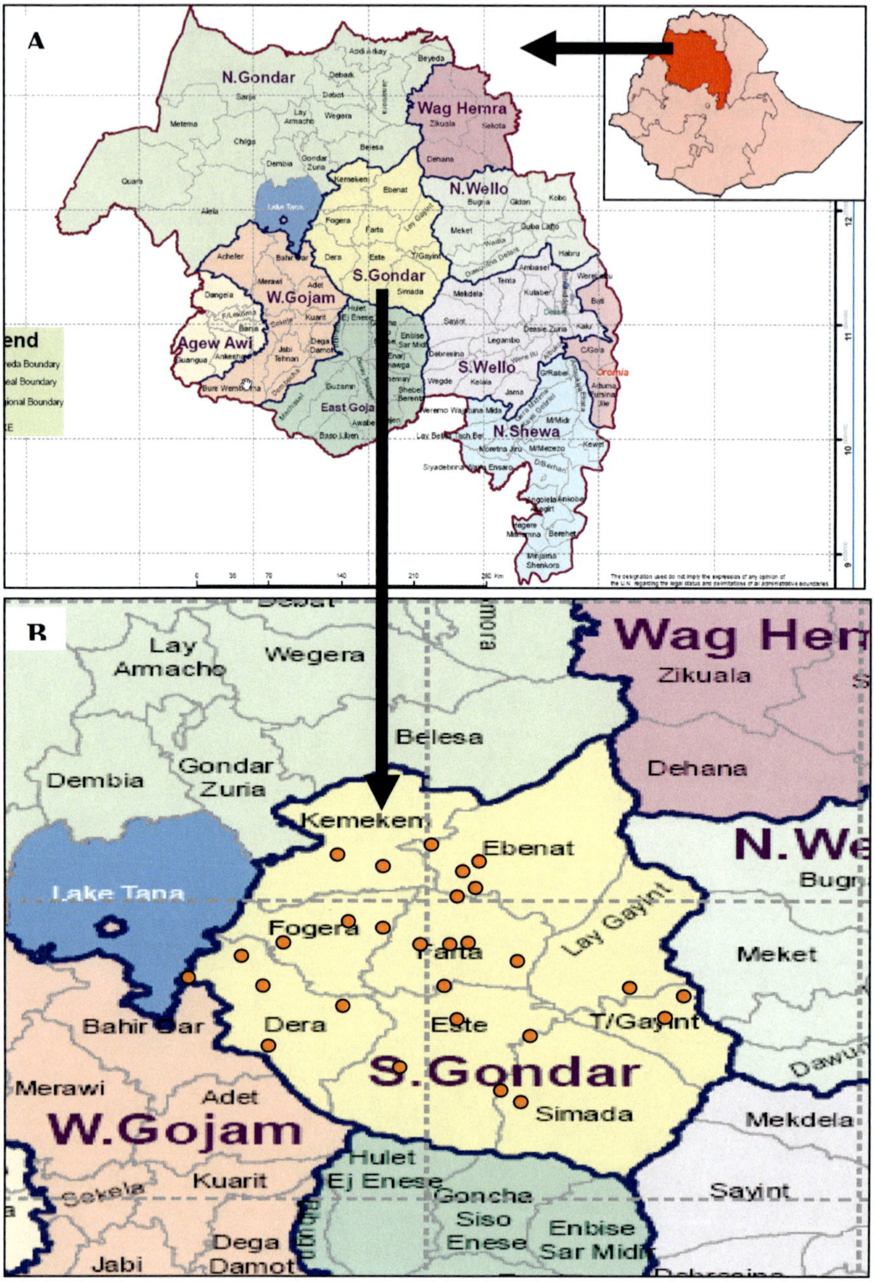

Figure 2. Map of study areas: A) The Amhara National Regional State. B) The South Gondar Administrative Zone (Scale 1 cm to 19 km). Dots show location of church forests studied.

Outline of the thesis

This thesis consists of seven chapters.

CHAPTER 1: introduces the background of the study and the Ethiopian Orthodox Tewahido Church and its tradition in conserving and preserving forests. It gives basic concepts of habitat fragmentation and ecological restoration in the Ethiopian context. It also highlights how to assess ecosystem function in remnant forests and describes major possible factors affecting regeneration in remnant forests. Finally it describes the overall objectives of the research and gives insight to the study area.

CHAPTER 2: addresses the question how and why forest structure, species composition and biodiversity vary across church forests. Species and structural composition is presented for 28 church forests. It shows the effects of altitude, forest area, cattle interference and wood harvest gradient on the forest structure, species composition and diversity. It also gives information about the role of altitudinal difference and geographical distance in determining similarity in species composition among church forests.

CHAPTER 3: addresses the question, do woody species maintain a soil seed bank in church forests? It describes the composition, density and spatial distribution of soil seed banks and its contribution for the regeneration of woody species in seven church forests. It discusses the results with the regeneration problems and implication of the future existence of the woody flora characteristic of church forests in particular and dry Afromontane areas in Ethiopia in general.

CHAPTER 4: addresses the question of how intense is post-dispersal seed predation in church forest and for how long tree seeds maintain their viability while buried in forest soil. It describes how seed predation varies with type of species and habitat in one church forest. It also reveals for how long tree seeds maintain their viability while buried in forest soil. It discusses the implication of post-dispersal seed predation and viability of seeds in forest soils for the regeneration of trees in church forest.

CHAPTER 5: asks how grazing affects the regeneration of indigenous tree species in church forests. It shows the effect of livestock grazing on seed germination, seedling survival and growth of four indigenous tree species in two church forests. The effects of grazing are compared between inside the forest and the open field. It gives information on whether or not the effect of grazing is species specific. It finally discusses the effect of livestock on the sustainability of church forests and their potential in the restoration of degraded landscape with practical recommendations.

CHAPTER 6: this chapter explores regeneration success (seedling establishment, seedling survival and growth of four selected tree species) along the gradient from the forest interior via the edge to the open field in one church forest. It also compares the effect of canopy gaps and closed canopy sites inside the forest. It tests the effect of different management interventions on regeneration, such as seed sowing, litter removal and weeding. It discusses which microsites are suitable for regeneration with their implication, and gives recommendations on interventions that can improve the regeneration of the forests.

CHAPTER 7: discusses the significance of church forest in Ethiopian forestry, their overall status and reveals major threats. It discusses major bottlenecks for regeneration at forest level and species level. For four species major bottlenecks are discussed at different life phases. It also indicates practical interventions to safeguard these forests and opportunities for restoration at landscape level.

CHAPTER 2

SPECIES AND STRUCTURAL COMPOSITION OF CHURCH FORESTS IN A FRAGMENTED LANDSCAPE OF NORTHERN ETHIOPIA

Alemayehu Wassie, Frank Sterck and Frans Bongers

Abstract

In Northern Ethiopia, while most of the dry afromontane forests have been converted to open agricultural lands, many small isolated fragments remain around churches ("church forests"). We assessed the species and structural composition of these forests and investigated whether and how this composition varies with altitude, forest area and human influence. 28 forests with a total of 500.8 ha were selected at different altitudes (range 1816 to 3111 masl) and of various sizes (range 1.6 to 100 ha).

A total of 168 woody species (100 tree species, 51 shrub and 17 climber) representing 69 families were recorded. Forests differed strongly in species number (15 to 78), basal area (4.8 to 111.5 m^2/ha), number of individuals ≥ 5 cm dbh (267 to 1553/ha), number of individuals >1cm diameter (619 to 2421/ha) and number of seedlings (0 to 5263/ha).

Basal area decreased with wood harvest but was independent from altitude, forest area and cattle interference. Species dominance increased with altitude and cattle interference. The ratio understory to upperstory density decreased with cattle interference but was independent of altitude and forest area. All species richness measures and diversity indices decreased with altitude, but were independent of forest area, cattle interference and wood harvest.

Similarity between forests decreased with altitude difference between forests, but geographical distance hardly explained variation.

We conclude that altitude is the main factor determining species composition while human influence determines structural composition of these forests. Particularly, cattle grazing strongly determines forest structure and species composition in the understory and is expected to have a strong longer-terms effect on whole forest structure and composition. Forest area has no significant effect on structural and species composition. This implies that although large size forests are a necessary element of successful conservation, small patches and appropriate matrix management could be useful complements.

Interconnecting these remnant forests by vegetation corridors following natural terrain or stream lines, creating buffer areas around them, excluding cattle interference, reducing intensity of wood harvest and developing more patches in the landscapes are possible landscape management activities. This will facilitate propagule and germplasm flow which may ultimately sustain these forests and help to restore the whole landscape.

1. Introduction

Worldwide, forests have been fragmented into small patches, and forest structure and species composition have been influenced due to this fragmentation and habitat loss (Echeverria, 2006). Particularly fragmentation of habitat has been described as one of the main causes of diminishing biological diversity in the tropics (Hill and Curran, 2001; Ross et al., 2002; Santos et al., 2007). Nonetheless most of the information available is from wet tropics while fragmentation effects in dry forests are rarely described. An extreme case of fragmentation and isolation is presented by remnant dry Afromontane forests in the northern Ethiopian highlands. Deforestation in Ethiopian highlands often involves the conversion of landscapes with continuous forest to many small remnant forest patches around churches and inaccessible areas set in a matrix of non forest vegetation (Demel Teketay, 1996; Bingelli et al., 2003; Alemayehu Wassie et al., 2005; Aerts et al., 2006a). Understanding the factors that influence patterns of species and structural composition in fragmented systems may be critical to conserving the remaining forests (Turner, 1996; Ross et. al, 2002). Here we explore how forest size, altitude and human activities influence the structure and species composition of church forests.

An important aspect of fragmentation is the change in forest area and its impact on species number and composition. Forest area reduction might lead to a decline in populations sizes via demographic or genetic events can influence species persistence (Shafer, 1981). Moreover, plant species reproduction and opportunities for expansion can be hampered due to interrupted or altered pollination and dispersal mechanisms, as well as migration patterns in the fragmented landscape (Aizen and Feinsinger, 1994; Benitez-Malvido, 1998). All these changes may result in shifts in plant species composition and structure. The changes tend to be more critical in smaller fragments than in larger fragments (Laurance and Yensen, 1991; Santos et al., 2007). However an overview of studies that have explicitly addressed the relationship between forest area and species composition in tropical forests found that some studies did find such a relationship while others did not (Debinski and Holt, 2000). The studies that did find relationships found that in general species richness, species abundance and species density increases with forest area. Several factors may distress the influence of area or blur the detection of an area effect alone. The inherent environmental heterogeneity among patches for instance due to altitude and the prevalence of disturbances can interact with area, influencing species composition and structure in forest fragments (Ross et al., 2002; Santos et al., 2007).

Many studies comparing altitudinal gradient of tropical forests showed that both species diversity and forest stature decrease with increasing altitude on tropical mountains (Lieberman et al., 1996; Aiba and Kitayama, 1999, Lacoul and Freedman, 2006) though there were also a few studies that reported no decline in species diversity with increased altitude (Lovett, 1996; Lovett, 1999). The low species diversity at higher altitudes may reveal the low rate of invasion and/or the high extinction rates of populations that colonize them (Stevens, 1992). The main explanations given were decreases in temperature, amount of photosynthetically active radiation and availability of nutrients, with increasing altitude (Grubb 1977; Aiba and Kitayama, 1999). However, as most studies were based on continuous forest, information about whether the trend is the same in fragmented forests is scarce.

Fragmentation in almost all times is associated to human induced disturbances which in turn influences community structure and maintenance of biological diversity (Soulé et al., 1992; Connell, 1978; Huston, 1979; Petraitis et al., 1989; Norton et al., 1995; Yates et al., 2000). The most common human induced disturbances are cattle interference and wood harvest. It is generally accepted that grazing up to intermediate levels of intensity can increase plant species diversity (Naveh and Whittaker, 1979) but unmanaged, intensive grazing coupled with wood harvest rather has a negative effect (Smiet, 1992; Ramirez-Marcial et al., 2001). For instance domestic grazing has affected most European forests for centuries, resulting in changed species composition of these forests (Bengtsson et al., 2000). Cattle interference could hamper regeneration and diminish the understorey structure and species composition. Understorey vegetation is the key element representing ecological changes that may occur a few years to decades later in the overstorey and entire forest. Wood harvest is mostly targeted on mature canopy trees but the process could have both negative and positive effect on the understorey. Comparison of understorey structure and composition to that of the overstorey could indicate best the effect of factors influencing the whole community (Dale et al., 2002; Onaindia, et al., 2004)

The effects of these factors on species and structural composition of forest remnants in the Ethiopian highlands are barely known. Here we analyse remnant forests around churches and monasteries in Northern Ethiopia. Ethiopian churches and monasteries have a long standing tradition of preserving and conserving their forests, including many native plants and animals (Alemayehu et al., 2005a,b; Bongers et al., 2006). This study assesses the species and structural composition of 28 church forests and investigates whether and how this composition varies with altitude, forest area and human influence. Specifically we address four questions: (1) what is the forest community structure and species composition of the church forests?; (2) How do altitude, forest area and human influence affect structure and species richness of these forests? For the second question we predict that (a) species richness and diversity decreases with increasing altitude, cattle interference and wood harvest but increases with forest area; (b) species dominance increases with increasing altitude, cattle interference and wood harvest but decreases with increasing forest area; (c) ratio of understorey to overstorey density decreases with cattle interference but increases with wood harvest; (d) ratio of understorey to overstorey species richness decreases with cattle interference and altitude but increases with wood harvest and forest area; (3) How do altitude difference and distance affect similarity in species composition among church forests? We hypothesized that similarity in species composition between two forests decreases with increasing distance and with altitude difference; and (4) Do altitudc, forest area and human influence affect similarity in species composition between understorey and overstorey of church forests? For this latter question we predict that similarity in species composition between understorey and overstorey of these forests decreases with increasing human influence and forest area but increases with altitude.

2. Material and Methods

2.1. Study Area

The study was conducted in South Gondar Administrative Zone (SGAZ), Amhara National Regional State, Ethiopia (Fig. 1, Map). Its geographical location is between 11° 02′ - 12° 33′ N and 37° 25′ - 38° 41′ E with an altitude range of 1500-4231 m. SGAZ has an area of about 14,299 km^2. The rainfall is characterized by a bimodal distribution with the major rainy season from June-August and short rainy season the 'Belg' from March-May. The annual average rainfall varies between 400 and 700 mm and the annual average temperature ranges between 9.3 and 23.7 ^{o}C. The topography comprises uneven and ragged mountainous highlands, extensive plains and also deep gorges. It is one of the most degraded and eroded areas in the regional state. The common soil types are Vertisol, Cambisol, Rigosol, Liptosols, Flovisols and Arenosols. The total forest cover in SGAZ is 20,882 ha, comprising 16,660 ha of natural forest and 4222 ha of man-made plantations (SGAZDA unpublished). The forest cover accounts for only 1.4% of the total area of SGAZ.

Out of the 1404 church forests found in SGAZ, 28 forests with a total of 500.8 ha forest were selected for the present study (Table 1). They were located at altitudes ranging from 1816 to 3111 m a.s.l., and had areas varying between 1.6 and 100 ha. The churches in these forests were established between 368 and 1984 A.D. The 28 forests fall in two agro-climatic zones of Ethiopia 'Woyna-Dega' (1500-2300 masl) and 'Dega' (2300-3200 masl).

2.2. Data Collection

For each church forest the forest area was calculated from GPS traverse readings. The area occupied by church buildings and open spaces were measured and subtracted from the total area to get the actual forest area. Within each forest the vegetation was censused in 10 x 10 m plots located at 50 m distance to each other along parallel transects. The first transect was aligned randomly at one side of the forest using a compass; then the others were laid at 50 m intervals from each other. Since the forest at Gelawdios was so extensive, transects and sample plots were laid at 100 m intervals. Due to difference in forest size, the number of plots varied between 7 to 56 (Table 1).

The mean canopy openness of the plots was estimated using Spherical Crown Densiometer. Leaf litter depth was measured from the litter surface to bare organic soil at 5 points, at the center and four corners of each plot. The number of clearly visible cattle trails and dead stumps (≥ 5 cm diameter) were recorded for each plot.

All woody plants within the sample plots were identified and recorded. Plant identification was done using (Hedberg and Edwards, 1989, Edwards, et al., 1995, Azene Bekele, 1993, Edwards, et al., 2000). For species that proved difficult to identify in the field, herbarium specimens were collected, dried properly and transported to the National Herbarium at Addis Abeba University for identification. For species found difficult to identify at herbarium a morphospecies code was given for the analysis.

Diameter at breast height (1.3 m, dbh) of all living woody plants (trees, shrubs, lianas) in the sample plots having ≥ 5 cm dbh were measured using diameter tape. Individuals with dbh < 5 cm diameter and > 1cm diameter at 10cm above ground were not measured but counted. Plants with multiple stems at 1.3 m height were treated as a single individual and the dbh of all the stems were taken. If a tree was buttressed and abnormal at 1.3 m, the diameter was measured just above the buttress where the stem assumes near cylindrical shape. Seedlings (here defined as woody plants with a diameter at 10cm above ground of < 1cm) were censused (identified and diameter measured at 10cm) in a sub plot of 5 x 5 m that was marked on the right front quarter of the main 10 x 10 m plot. The heights of all woody species were measured using a hypsometer and meter tape.

2.3. Data Analysis

Forest structural composition

Forest structure was described in terms of mean maximum height (calculated from the maximum height recorded in each sample plot), mean basal area (calculated from individuals with dbh ≥5cm), dominance, ratio of understorey to overstorey density and size class distribution. Selected structural characteristics of the forests (canopy cover, maximum height, mean number of individuals/ha (seedlings/ha, >1 cm diameter at 10cm height and ≥ 5cm dbh), basal area/ha, and litter layer thickness were analysed for cross character correlationships (Pearson's r). Main factors forest area, altitude and human influence were checked for independency. Human influence in all analyses here after was characterized by two aspects: cattle interference (mean number of cattle trails/plot) and wood harvest intensity (mean number of dead stumps/ha).

Dominance is described in terms of important value index (IVI) which is the sum of relative basal area, relative density/abundance and relative frequency/occurrence in the sample plots for each species (Curtis and McIntosh, 1950; Pascal and Pelissier, 1996). IVI may vary between 0 and 300% but was converted into a 100 percent scale. Ratio of understorey to overstorey density for each forest is calculated as the number of seedlings (individuals <1cm diameter at 10 cm height) divided by the number of individuals ≥ 5cm dbh, averaged over the plots. Hereafter understorey refers to all individuals <1cm diameter at 10 cm height (the seedlings described above) while the overstorey refers to all individuals≥ 5cm dbh. The ratios as dependent variable were regressed with altitude (km), forest area (ha), cattle interference and wood harvest as independent variables. The highest IVI, the sum of the first highest five, the sum of the first highest 10 IVI and ratios from each forest were regressed with altitude (km), forest area (ha), cattle interference and wood harvest as independent variables.

For all forests tree size class distributions were formed, based on tree diameters. The 28 forests were grouped into four size class distribution patterns. Difference between groups with respect to altitude, forest area, cattle interference and wood harvest were analyzed using one way ANOVA. The percentage of non-tree woody species in each class was also calculated to quantify the contribution of life forms in the structure of the community.

Species richness and diversity

As a measure of species richness, the number of observed species in a community or sample, and also two non-parametric estimators of species richness, the Chao1 and the first order jackknife were used (Gimaret-Carpentier et al., 1998; Krebs 1999; Magurran 2004). The Cole rarefied number of species was calculated based on the smallest total number of individuals encountered (98, at Ascha) (Sanders, 1968). For comparisons among forests of the increase in species numbers with increase of number of plots or number of individuals, mean (n=50) randomized species accumulation curves for both sample plots and individuals were constructed for each forest. Additionally a variety of commonly used diversity measures, i.e. Shannon-Wiener's, Simpson's and Fisher's diversity indices were computed (Magurran 2004).

Ratio of understorey to overstorey species richness for each forest is calculated as the rarefied number of species in the understorey divided by the rarefied number of species in the overstorey based on the smallest sample number of plots (7, at

Ascha). Species richness measures, diversity indices and ratio of understorey to overstorey species richness as dependent variable were regressed with altitude (km), area (ha), cattle interference and wood harvest as independent variables. To meet normality assumption species richness measures and diversity indices were log-transformed except for Shannon-Wiener's index.

Species composition similarity

Similarity between forests and between understorey and overstorey in species composition was analyzed using Chao-Jaccard abundance-based similarity coefficient (Sj) following Chao et al. (2005). The range of this coefficient varies from 0 (no similarity) to 1 (complete similarity).

Similarity between forests was regressed on altitude difference (km) and distance (km) separately. As the regression function for similarity and altitude showed wider scatter at lower altitude than at higher altitude a quantile regression was performed using Blossom Statistical Software version 2005 (Cade and Richards, 2005). Altitude and distance were also fit in a multi linear regression model as independent variables with similarity coefficient as dependent variable. Similarity between understorey and overstorey as dependent variable were regressed with factors altitude (km), area (ha), cattle interference and wood harvest.

All species richness and diversity indices as well as similarity of species composition were analyzed using EstimateS 8.0 statistical software (Colwell, 2005) and rarefaction of species richness using EcoSim 5.0 software (Gotelli and Entsminger, 2004). All regressions and correlations were analyzed using SPSS 12.0 (Field, 2005).

3. Result

Overall species and structural composition

A total of 168 woody species (100 tree species, 51 shrubs and 17 lianas) representing 69 families were recorded in the 28 church forests studied (Appendix 1). Out of these species, 160 were indigenous and only eight were exotic (2 shrub and 6 tree species) representing 6 families. The families with the highest number of species were Fabaceae (17), Euphorbiaceae (10), Moraceae (10) and Rubiaceae (7). 37 families were represented by a single species only. The total number of species and families in each of the 28 church forests ranged from 15 and 12 at Wonberoch to 78 and 44 at Wonkshit, respectively (Table 2). *Juniperus procera* L., *Olea europaea* L. and *Maytenus arbutifolia* (A. Rich.) Wilczek were the most important species in terms of IVI in all forests pooled (Annex 1). 46 species had extreme low overall IVI (IVI<1, all 28 forests pooled) and 53 species were found only in one forest.

Forests varied widely in their structural characteristics (Table 2). Mean basal area ranged from 4.8 m^2/ha to 111.5 m^2/ha. Mean number of individuals ≥ 5 cm dbh ranged from 267/ha to 1553/ha, mean number of individuals >1cm diameter ranged from 619/ha to 2421/ha and mean number of seedlings ranged between 0 and 5263/ha. The average maximum height ranged between 9.1 m and 22.9 m. Mean canopy cover ranged from 18% to 86% and litter thickness from 0.5 cm to 2.51 cm. The highest cattle trail was 4.4/plot and lowest was no cattle trail. The highest mean number of dead stumps recorded was 149/ha and the lowest 14/ha.

Canopy cover significantly correlated with basal area (Pearson correlation: $r=0.47$, $p=0.01$), number of stems (≥5cm dbh) ($r=0.58$, $p=0.001$) and height ($r=0.59$, $p=0.001$). Height also positively correlated with basal area ($r=0.48$, $p=0.01$). Forests with higher number of stem (≥5cm dbh), basal area and height tend to have higher canopy cover. Litter layer thickness increases with canopy cover ($r=0.40$, $p=0.03$), basal area ($r=0.4$, $p=0.02$) and total density (> 1cm diameter) ($r=0.61$, $p=0.001$) and decreases with cattle interference ($r=-0.63$, $p=0.001$) and IVI of dominant species ($r=-0.55$, $p=0.002$). There was no significant correlation among altitude, forest area and human influence (Pearson correlation, $r< |0.1|$, $p>0.4$) showing that these factors were independent to each other. However, the two aspects of human influence, cattle interference and wood harvest were positively correlated ($r=0.60$, $p<0.001$).

Structural composition

Basal area decreased with wood harvest (r=-0.39, p=0.04) but was independent of altitude, forest area and cattle interference (p>0.05). Mean maximum height was not significantly correlated to altitude, forest area, cattle interference or wood harvest (p >0.05).

The dominance of the most abundant species increased significantly with altitude (all three variables) and cattle interference (only the first most dominant species) (Fig 2 A, C & Table 4). However, it was independent of forest area and wood harvest intensity (Fig2 B, D, $r^2<0.06$ & p>0.05).

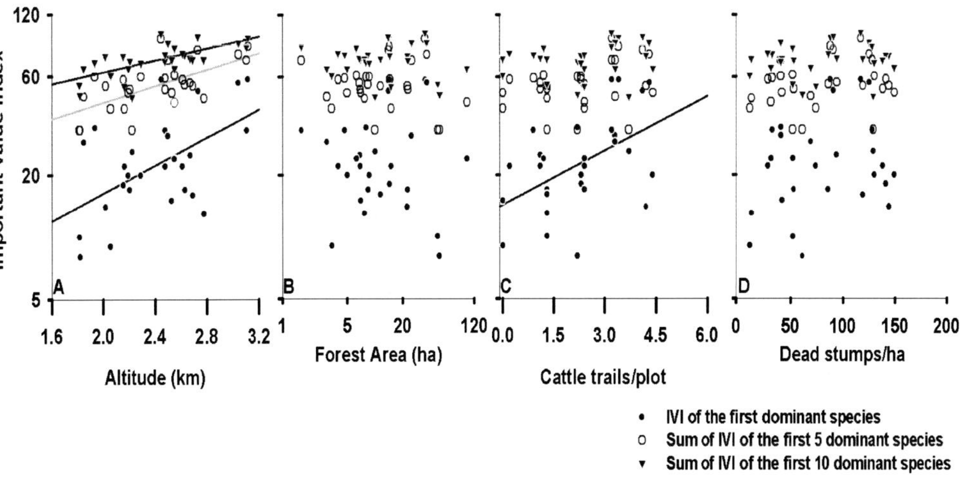

Figure 2. Species dominance in terms of important value index (IVI) as a function of altitude (A), forest area (B), cattle interference (C) and wood harvest (D). All values on y-axis and forest area are in log scale. Three levels of dominance are given, the IVI of the first dominant species (filled circles), the cumulative IVI of the first 5 (open circles) and that of the first 10 (triangles) most dominant species. The lines give significant linear regressions (Formulas and r^2 of the lines are presented in Table 4).)

The understorey to overstorey density ratio decreased with cattle interference (Fig.3 C3, Table 4) but was independent of altitude and forest area (Fig. 3 C1,C2, $r^2<0.04$, p>0.05). The decrease is most probably due to the effect on seedling density as mean seedling density strongly decreased with cattle interference (r=-0.62, p=0.000). The relationship with wood harvest however exhibited a quadratic pattern: for a low wood harvest the ratio decreases but for a high wood harvest the ratio increased (Fig. 3 C4, Table 4). This was mainly the result of the outlying effect of two sites (Woji and Alember). During field work we observed that in Woji many trees were harvested for construction and in Alember all large plants were cleared (for safety reasons) in a band 50 mt wide along a hydroelectric power line passing over this forest.

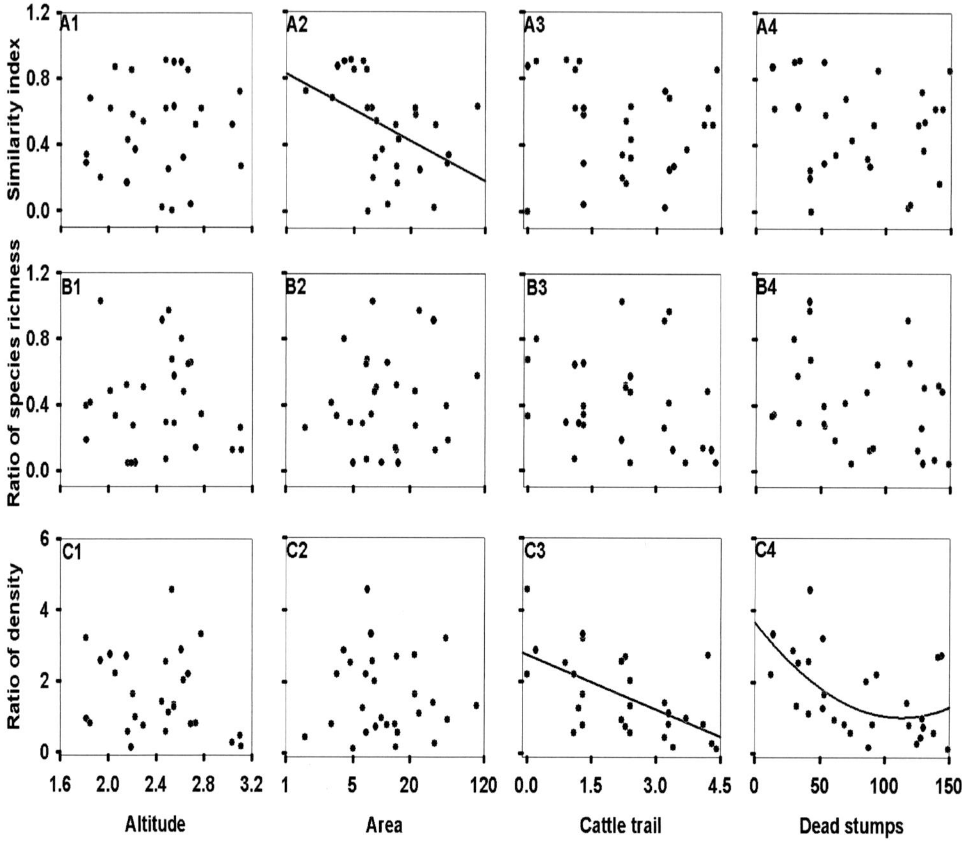

Figure 3. Species similarity (A1-4), ratio of understorey to overstorey rarefied species richness (B1-4) and ratio of understorey to overstorey density (C1-4) as a function of altitude, forest area, cattle interference and wood harvest. Forest area is given on a logarithmic scale. (Formulas and r^2 of the lines are presented in Table 4).

The tree size class distributions of the 28 forests were grouped into four patterns of distribution (Fig. 4). The first pattern (group I, 8 forests) is characterized by a large number of individuals in the first class, descending towards the higher size classes (more or less inverted J shape). The second pattern (group II, 6 forests) is characterized by a high number of individuals in the first and last classes and lower number in the intermediate ones (U shaped curve). The third pattern (group III. 4

forests) is characterized by a high and equal number of individuals in the lowest 4-5 classes, decreasing to the larger classes. The fourth pattern (group IV, 10 forests) has few individuals in the first class and many individuals in the 2nd to 5th size class. The four groups were significantly different from each other in the number of dead stumps ($F_{3,\,24}$=4.9, p=0.008) and cattle interference ($F_{3,\,24}$=5.0, p=0.007) but not in their altitude ($F_{3,\,24}$=1.1, p=0.36) or forest area ($F_{3,\,24}$=1.2, p=0.31). Forests in Group I had lowest cattle interference (1.0 cattle trail/plot) and forests in Group IV had highest (3.6 cattle trail/plot) while others in between (forests in Group II had1.9 and Group III 2.8 cattle trail/plot). Forests in Group I had also the lowest number of dead stumps (31.6 dead stumps/ha). Group IV forests had the highest (109.5 dead stumps/ha) but this was not significantly different from group II and III (p=0.33 and 0.96 respectively). Human interference thus is the most important factor determining the size class distribution of the forests.

In all 28 forests in the first size class (0-5 cm dbh) of the community of woody plants the contribution of trees was less than 50% (47-10%) except for one forest (Deder, 63%). In other words the first size class of the community is dominated by non tree forms of woody species (Fig. 4).

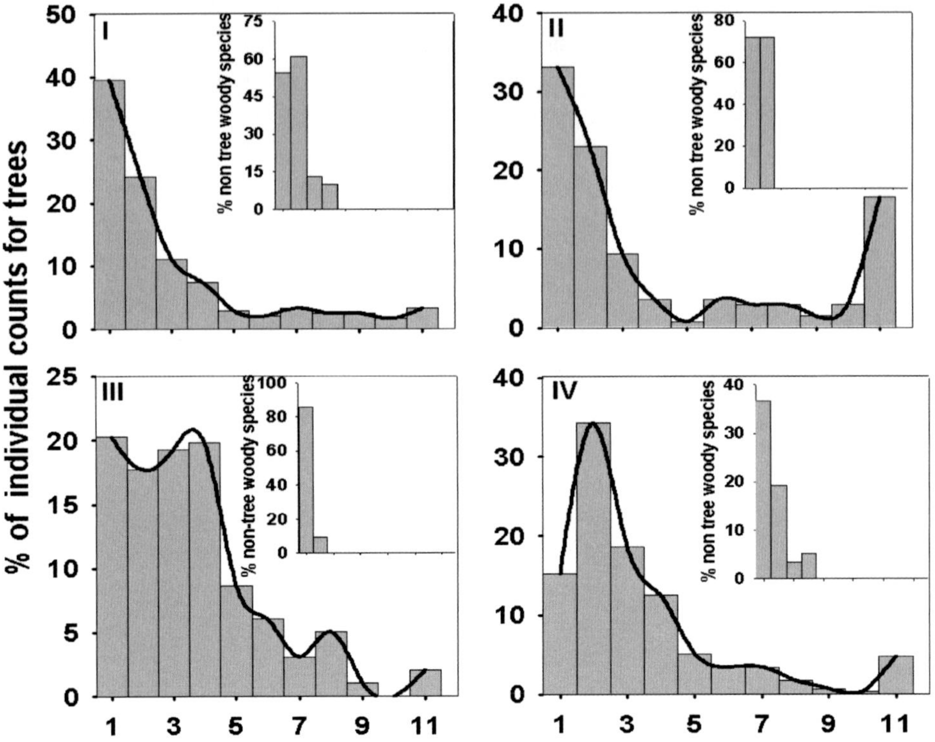

Figure 4. Size class distribution of trees (main figures) and contribution of non-tree woody plants in the first size class (inset figures) for the 28 forests, grouped into four major types. Diameter class for both main and inset figures 1: 0-5; 2: 5-10; 3: 10-15; 4: 15-20; 5 :20-25; 6: 25-30; 7: 30-35; 8: 35-40; 9: 40-45; 10: 45-50; 11:>50 cm dbh.

Species Richness and Diversity

Forests varied widely in their species richness and diversity (Table 3, Fig. 5). Observed number of species ranged from 15 to 78, while the rarefied number of species based on 98 individuals ranged from 10 to 45. Both Chao's and Jackknife's Indices estimated higher species richness for all forests as compared to the observed number of species. According to Chao's and Jackknife's estimates the highest number of species was 99 and 100 respectively and the lowest were 17 and 19. The species area curves (Fig. 5) also shows increasing trend as the number of sample plots added suggesting that large sample sizes could bring more number of species in these forests. Fisher-alpha diversity ranged between 3 and 31, Shannon-Wiener from 1.6 to 3.8 and Simpson from 3.5 to 41.5. The species richness measures (observed, rarefied, Chao's and Jackknife's estimates) were strongly correlated (Pearson r= 0.74-0.97 ; p=0.000) and so were the three diversity indices (Pearson r= 0.86-0.88 ; p=0.000). All species richness and diversity values decreased with altitude (Table 3), but were independent of forest area, cattle interference and wood harvest ($r^2<0.04$ and $P>0.05$).

Observed and rarefied number of species of the entire forest (Fig. 6 A1-2) and of the overstorey (≥5cm dbh) (Fig. 6 B1-2) decreased with altitude. However, understorey species richness (seedlings) was not affected by altitude alone, but rather the interaction effect of altitude and cattle interference showed a significant effect. Both the observed and rarefied number of species in the understorey decreased with both altitude and cattle interference (Fig. 6C1-2). On the other hand, the ratio of understorey to overstorey species richness was not significantly related to any of the factors considered (altitude, forest area, cattle interference or wood harvest, Fig. 3 B1-4 ; $r^2<0.02$, $p>0.05$).

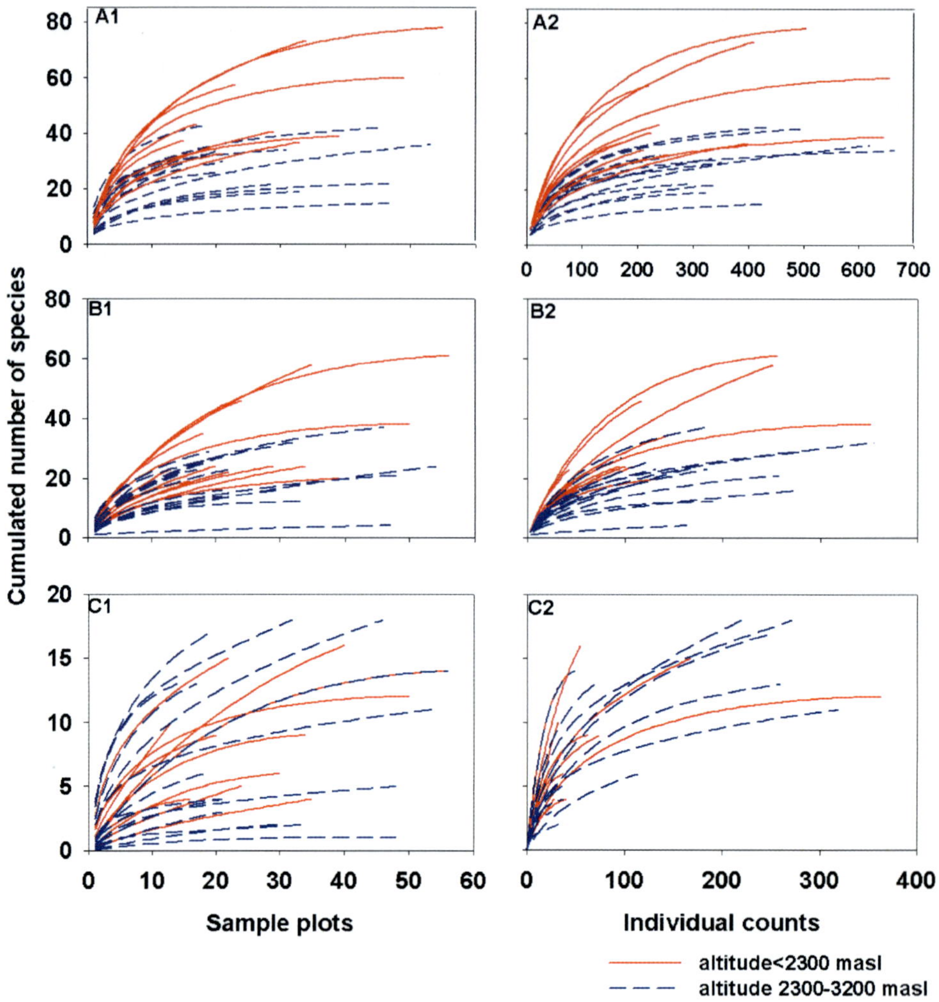

Figure 5. Randomized mean number of species cumulated for entire forest (A1-2), overstorey (B1-2) and understorey (C1-2) versus number of sample plots and individual counts for the 28 forests.

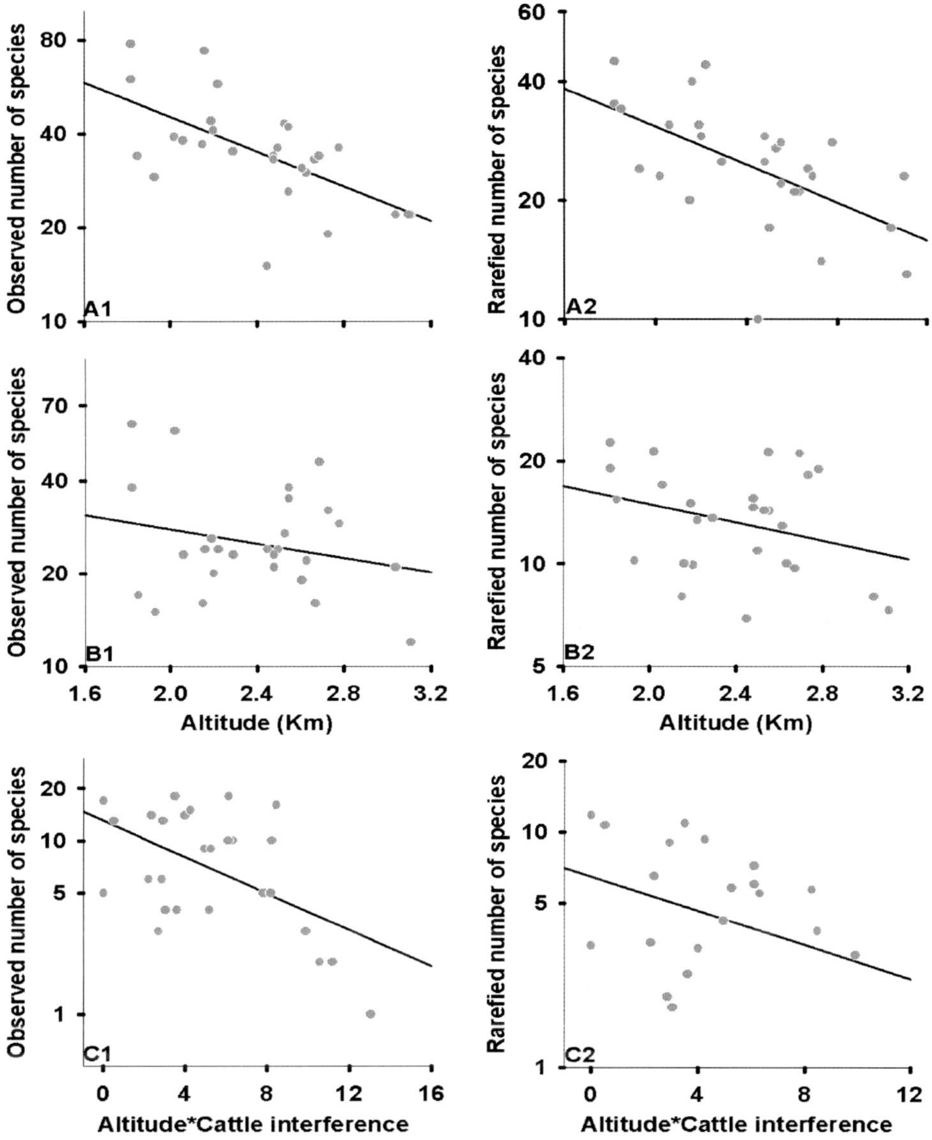

Figure 6. Observed and Cole-rarefied species numbers for 28 forests. Values for the entire forest (A1-2) and the overstorey only (B1-2) as a function of altitude. Values for the understorey (C1-2) as a function of interaction value between altitude and cattle interference. (Formulas and r^2 of the lines are presented in Table 4).

Species composition similarity

Similarity in species composition between two forests decreases as altitude difference and distance between them increases (Multiple regression: r^2 = 0.36; β_o = 0.58, p = 0.000; coefficient for altitude= -0.52, p = 0.000; coefficient for distance = -0.002, p = 0.000 and coefficient for interaction effect (altitude difference X distance) =0.003, p=0.003).

When we partitioned and computed independent regression similarity between forests decreased with altitude difference between the forests and explain most the variations as the full model (r^2 =0.34), but geographical distance in itself hardly (r^2 only 5%) explains similarity (Fig. 7, Table 4). The highest species similarity (0.83, Jaccard abundance based coefficient) was found between Ascha and Deder which are located at 11 m altitude difference and 31 km distance from each other. The lowest species similarity 0.02 was between Zhara and Ascha, located at 1166 m altitude difference and 61 km from each other. The effect of altitude difference is especially strong when maximum similarity is taken into account (95^{th}quantile, Fig. 7A). Two forests with low altitude difference do not necessarily have high similarity in species composition, however (points below 5^{th}quantile, Fig7A).

Species composition similarity between understorey and overstorey decreased with forest area but had no relation with altitude, cattle interference and wood harvest (Fig.3, A1-4). In larger forests the similarity between two stories decrease while in smaller forests similarity tends to be higher.

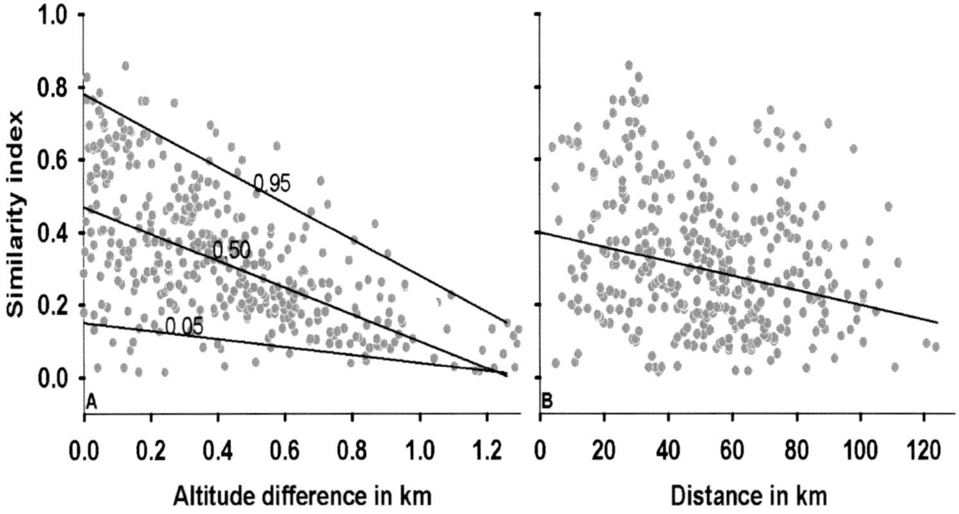

Figure 7. Similarity between forests as a function of (A) difference in altitude and (B) distance. The three lines in (A) show the 5th, 50th & 95th quantiles, the line in B shows the 50th quantile. (Formulas and r^2 of the 50th quantile lines are presented in Table 4).

4. Discussion

Overall floristic composition

Many studies suggested that the highlands of Ethiopia were once covered by diverse forest dominated by *Juniperus procera,* co-dominated by *Olea*, presumed to be the natural vegetation of the region (Logan, 1946; Friis, 1992; Demel Teketay, 1992). Darbyshire et al. (2003), using pollen and charcoal analysis, further substantiate that since AD 1400-1700 *Juniperus* forests with *Olea* and *Celtis* had been the predominant vegetation of the Northern Highlands of Ethiopia. Our church forests were co-dominated by these species suggesting that these forests represent remnants of the devoid natural vegetation of the Northern Ethiopia. Soil seed bank assessment on seven of these 28 forests (Alemayehu Wassie and Demel Teketay, 2006; Chapter 3) showed that among the woody species Juniperus had the highest proportion of seeds and also their high viability may indicate that this species possesses a potential for forming soil seed banks, which help to ensure its perpetuation in the events of disturbances. *Olea* compared with its slow growth has high survival success once established due to a strong tradeoff under environmental stress (Bacelar et al., 2006). Moreover its seeds dispersability by birds from one patch to another over the degraded matrix (Aerts et al., 2006b) may have helped the species to appear widely in these forests. Apart from ecological and climatic reasons the social values of these two species can contribute to their dominance. Almost all local churches are built from Juniperus and Olea woods. Most local people consider *Juniperus* as a tree that should belong to the church forest for its grace, strength and durability. In some areas we observe that some *Juniperus* trees left on farmlands still were considered as part of the church forest while this doesn't apply for other species. *Maytenus arbutifolia,* the third important species, has a wide distribution in Ethiopia. This shrub species is more abundant after disturbance (Tadesse Woldemariam, 1998) and hence may indicate those forests have been intruded by anthropogenic factors.

Unpublished records of the South Gondar zonal department of agriculture (SGAZDA unpublished) claim there are 125 woody species in the study area, but in our 28 church forests covering only 500 ha a total of 168 species are found. Moreover, our species area curves show increasing trends suggesting that we missed a considerable number of species for these forests. The fact that we only surveyed 28 churches out of 1404 churches available in the study area strongly suggests that it is most likely that more species can be found if more church forests are assessed. This may show that the species diversity of the area is not exhaustively identified and the potential of these church forests have been also overlooked. The overall species richness of these remnant forests was higher than that of Wof-Washa forest (51 woody species) (Demel Teketay and Tamirat Bekele, 1995), which is the biggest continuous forest in central highlands of Ethiopia. The maximum basal area/ha ($111 m^2$/ha) recorded in our forest sample is comparable to that of Wof-Washa forest (100.3 m^2/ha) (Tamirat Bekele 1994). The minimum basal area /ha (4.8 m^2/ha), number of individuals ≥ 5 cm dbh (267/ha) and canopy cover (18%) recorded were from same forest (Asolit) where the church was built very recently (1984) and the forest has been developing since then (Table 1). Based on historical records (Melaku Bekele, 1998) generally most of the forests which are associated with ancient churches appeared as left over of the original high forest while the recent ones are mostly rehabilitated wood lands after church establishment.

Forest area gradient

We expected that diversity would increase with forest area, but found no proof for that. Only the similarity in species composition between understorey and overstorey decreased with increasing forest area. Possibly the larger the forest area, the larger the pool for seedlings and mature individuals available which in turn decrease the chance of similarity between the under and overstorey. All other analysed variables (basal area, species richness, species diversity, species dominance, ratio of understorey's to overstorey's density, size class distribution) were not affected. Absence of a relationship between fragment area and species composition was also reported by other studies (Debinski and Holt, 2000; Haig et al., 2000; Santos et al., 2007). While large patches are needed by many species to maintain viable populations, it is important to recognize the complementary value of small remnants (Fischer and Lindenmayer, 2001). On the other hand, the effects of forest area need time. As our result is based on spatial variation only, we may not completely disregard forest area effects on diversity and structure of forests (Debinski and Holt, 2000).

Altitudinal gradient

At higher altitude, forests are less diverse and the dominance by a single to few species is strong. Similarly many studies in the tropics showed that species richness and diversity approximately linearly decrease with altitude (Stevens, 1992; Gentry, 1988; Aiba and Kitayama, 1999; Lacoul and Freedman, 2006). Some studies however, showed a unimodal species richness response to altitude. For instance

Grytnes and Vetaas (2002) showed an optimum between 1000-1500 masl in tropical Nepal. Although we cannot exclude that a unimodal trend may hold true in Ethiopia, at least within the altitudinal range we studied (1800-3200 masl) the trend of linearly decreasing species richness and diversity with altitude is clear. In our study, the species richness in the overstorey showed a similar trend as the entire forest but the understorey was affected by the interactive effect of altitude and cattle interference. The numbers of seedlings were affected by cattle grazing which mainly constitute understorey. Thus, the negative effect of cattle on the species richness of the understorey was much stronger at higher altitudes.

The effect of altitude was also important for similarity in species composition between forests. Maximum similarity in species composition was found with minimum altitude difference between forests. Geographical distance had only a weak effect on similarity although some studies reported that geographical proximity was a key factor for similarity in species composition (Santos et al., 2007). Vazquez and Givnish (1998) found a similar trend in tropical forests of Mexico: similarity in species composition declines with altitude difference between samples. Furthermore they also found that horizontal distance is about 17% as effective as vertical distance in fostering compositional divergence of forest species. The weaker effect of distance on similarity may indicate two possibilities. Either there had been no species composition gradient over the whole landscape (Tuomisto et al., 2003), except with altitude before fragmentation or these forests have been very isolated and stopped interaction between them due to the complex matrix. If the latter is true, forests in short distance couldn't exchange propagules and germplasms and as a result are developing a different species composition.

On the other hand at lower altitude difference we noticed that similarity was in some cases low. The total variation explained by altitude difference and geographical distance was only 36 %. This may indicate that there are also other factors that determine similarity in species composition between two forests like edaphic, topographic features, management, microclimatic factors and time of isolation (Gentry, 1988; Aiba and Kitayama, 1999; Ross et al., 2002; Santos et al., 2007).

Human influence gradient

Human influence had a strong effect on structural composition of the forest, but not on species richness and diversity except on understorey species richness. The inverse relation between basal area and number of cut stumps suggests a negative effect of wood harvesting on standing stock. The most exploited tree species we observed in the higher altitudes were the most dominant species (*Juniperus* and *Olea*), which are known for their valuable woods. In one of the forests (Woji, 22.5 ha) for instance, we counted 600 high canopy *Juniperus* trees cut just while we made the survey. It is interesting that even if these two species are the most exploited ones, still they tend to dominate these diverse forests. The wood is harvested for church building construction and maintenance, but we cannot exclude other uses. At lower altitudes *Acacia* and *Combretum* species were most exploited, mainly for charcoal and farm tools. In line to our findings, Pueyo et al. (2006) reported that community structure analysis proved more sensitive than species diversity indices to grazing effects. Onaindia et al. (2004) also showed that plant species richness did not vary with disturbance

Dominance increased with cattle interference. Some well palatable species are expected to be more severely affected by cattle than others, perhaps in the long run resulting in dominance of grazing resistant species (McNaughton, 1983, 1985; Pueyo et al., 2006). The ratio of understorey to overstorey density decreased with cattle interference, as expected. A light wood harvest has also diminishing effect on the ratio due to its damaging effect on seedlings during harvesting. With intense harvesting, however, more space (and light) is available for seedlings to germinate and develop. The size class distribution was also determined mainly by cattle interference and wood harvest as the four patterns (Fig. 4) were different to each other in these two factors. The lower percentage of the smallest size class in group IV (Fig. 4) was due to higher cattle interference. The typical inverted J shape of Group I, normally interpreted as a healthy size class distribution (Peters, 1996), was the result of a low cattle interference and a light wood harvest. This may indicate that human influences cause the variation in size class distribution among the forests. The larger share of non-tree forms of woody species in the first class of the community structure could imply tree species had lower recruitment probabilities as compared to lianas and shrubs in these forests. Lianas and shrubs may be better resistant to cattle grazing and disturbances than tree seedlings. Lianas and shrubs do have more ability to propagate vegetativly than tree species in tropical climate (Greig, 1993). It is also reported that vegetative propagation is particular characteristics to certain habitats such as areas subjected to grazing (Abrahamson 1980). Therefore, the future floristic composition of these forests may alter to bushy and shrub land as human interference continued.

Implications for conservation and restoration

In our forests altitude is the main determinant of species composition and human influence determines structural composition of these forests. In the understorey the interaction effect of altitude with human influence (particularly cattle interference) determines the species composition. Not only the species composition but also the number of seedlings were severely affected by cattle interference and few, if any, seedlings were found in some of the forests. In the long run this would probably lead to an human effect on overstorey composition of these forests as well. Forest area has hardly any significant effect on the number of species per area but the number of species is small in smaller forest. This implies that although large sized forests are a necessary element of successful reproduction (Shafer, 1981; Laurance and Yensen, 1991; Santos et al., 2007), small patches and appropriate matrix management could be useful complements for biodiversity conservation (Margules and Pressey, 2000; Fischer and Lindenmayer, 2001). Moreover, not only forest area but selection of remnants across a geographical variation, in this case mainly altitudinal variation, is important in maximizing species diversity (Hill and Curran, 2001). There are other factors that possibly determine species and structural composition of these forests like edaphic factors, topographic features and microclimatic factors. These should be given due attention in future studies.

Species richness estimates (Chao's 1 and Jackknife's 1) for all churches is higher than the observed number of species, which indicates that there are many species represented by single individuals (rare species) and also many species found only in single plots (unique species) (Table 3). This presence of many rare and unique species, makes these forests sensitive for conservation, as they are the only remnant forests in the area and the last option to hold those woody species. The possible explanation for presence of many rare and unique species could be attributed to the continuous reduction of forest area and poor dispersability of the species.

Interconnecting these remnant forests by vegetation corridors following natural terrain or stream lines, creating buffer areas around them, excluding cattle interference, reducing intensity of wood harvest and developing more patches in the landscapes are the possible matrix management to facilitate propagules and germplasm flow and ultimately sustain these forests and restore the whole landscape. The pollen records of Darbyshire et al. (2003) show that forests in northern Ethiopia can regenerate after as much as 1800 years (500 bc to ad 1300) of anthropogenic clearance. Their results strongly suggest that forest cover in the area could again increase in the future, under appropriate land management. How fast forest can regenerate and what kind of forest we expect mainly depend on the extent of fragmented forests available along the landscape (Cunningham and Saigo, 1995; Laurance and Bierregaard, 1997; MacMahon, 1997; Eshetu Yierdaw, 2002; SER, 2002). These forest fragments are stepping stones and building material for restoration of these afromontane landscapes.

Table 1. Characteristics of the 28 church forests selected for this study.

Name of The Church	Altitude (M.a.s.l)	Forest Area (Ha)	No. of Plots *	Year of Estab. of the Church	District/ Woreda	GPS Location
Deder	3111	14.4	30	1690	Simada	N11^{o} 32' & E38^{o} 14'
Ascha	3100	1.6	7	1940	Laygaynt	N11^{o} 43' & E38^{o} 28'
Dedim	3037	36.6	48	1330	Farta	N11^{o} 47' & E38^{o} 12'
Asolit	2777	7.7	21	1984	Tachgaynt	N11^{o} 41' & E38^{o} 35'
Abegeldi	2731	14	34	1735	Ebenat	N12^{o} 09' & E38^{o} 07'
Debresena	2690	11.5	32	1570	Farta	N11^{o} 51' & E37^{o} 59'
Zagua	2669	6.8	17	1450	Farta	N11^{o} 42' & E38^{o} 02'
Wuahir	2628	8.4	22	1360	Farta	N11^{o} 51' & E38^{o} 05'
Hiruy	2611	4	14	360	Farta	N11^{o} 51' & E38^{o} 03'
Enshet Kuskuam	2550	6.3	21	1410	Tachgaynt	N11^{o} 37' & E38^{o} 32'
Gelawdios	2549	100	46	1500	Dera	N11^{o} 38' & E37^{o} 48'
Mosha	2529	7	19	1320	Estie	N11^{o} 26' & E37^{o} 56'
Dengolt	2500	25	54	1300	Estie	N11^{o} 36' & E38^{o} 04'
Mekedesemariam	2478	4.7	18	1430	Simada	N11^{o} 19' & E38^{o} 12'
Gunaguna	2477	6.9	21	1260	Ebenat	N12^{o} 03' & E38^{o} 06'
Wonberoch	2447	35	48	1740	Ebenat	N12 ^{o}11' & E38^{o} 00'
Gibtsawit	2290	8.7	20	1620	Ebenat	N12^{o} 07' & E38^{o} 05'
Shamo	2222	10	24	1270	Libo	N12^{o} 07' & E37^{o} 53'
Amstya	2204	22.8	30	1260	Ebenat	N12^{o} 01' & E38^{o} 04'
Mantogera	2192	5	18	1275	Libo	N12^{o} 09' & E37^{o} 47'
Gedamselase	2161	15	35	1011	Simada	N11^{o} 21' & E38^{o} 10'
Alember	2153	14.5	34	1702	Fogera	N11^{o} 54' & E37^{o} 53'
Emashenkor	2056	3.4	16	1659	Dera	N11^{o} 42' & E37^{o} 37'
Woji	2017	22.5	40	1695	Fogera	N11^{o} 55' & E37^{o} 48'
Zhara	1934	8	22	1550	Dera	N11^{o} 48' & E37^{o} 34'
Quar	1850	3	13	1480	Fogera	N11^{o} 51' & E37^{o} 40'
Wonkesht	1820	50	56	1500	Dera	N11^{o} 30' & E37^{o} 37'
Qorata	1816	48	50	1655	Dera	N11^{o} 45' & E37^{o} 26'

* each plot measures 10x10 m

Table 2. Area of forests (ha), number of species and plant families, Mean basal area (m^2/ha, calculated form individuals with dbh ≥5cm), mean density (number of individuals /ha), average maximum height, mean canopy cover, mean litter layer thickness, mean number of dead stumps/ha (≥5cm diameter), mean number of cattle trails/plot and IVI found in the 28 church forests.

Name of The Church	Area of Forest	Species	Family	Basal Area M^2/Ha	Density /Ha ≥5cm dbh	Density /Ha 1-5cm dbh	Density /Ha Seedlings	Maxim. Height	Canopy Cover %	Litter Layer cm	Dead Stumps /Ha	Cattle trails /plot	Important Value index (IVI) Top1	IVI Top5	IVI Top10
Gelawdios	100	42	29	52.7	407	1109	476	17.1	84	2.51	32	2.4	24	45	66
Wonkesht	50	78	44	34.2	455	920	350	16.4	74	1.21	61	2.2	8	33	49
Qorata	48	60	32	68.2	704	1368	2904	16.2	66	2.30	52	1.3	10	33	55
Dedim	36.6	22	19	35.0	538	619	25	16.0	76	0.54	125	4.3	56	77	88
Wonberoch	35	15	12	13.9	340	910	308	9.4	42	0.54	117	3.2	58	92	98
Dengolt	25	36	28	35.8	326	1169	590	15.4	75	0.52	41	3.3	31	72	88
Amstya	22.8	41	30	32.7	363	773	493	17.1	59	1.52	53	1.3	17	52	72
Woji	22.5	39	24	25.0	348	1650	550	9.4	47	1.49	144	4.2	14	54	76
Gedamselase	15	74	41	34.8	717	1203	320	9.1	49	0.91	74	2.4	22	42	54
Alember	14.5	37	27	30.0	294	1206	729	14.5	59	1.56	141	2.3	18	58	76
Deder	14.4	22	19	70.0	920	1163	93	14.8	74	0.50	88	3.4	58	84	93
Abegeldi	14	19	17	21.6	559	979	376	12.7	64	0.62	91	4.1	51	81	94
Debresena	11.5	34	27	46.4	1110	2141	847	13.6	86	2.34	119	1.3	16	54	73
Shamo	10	58	37	28.2	488	954	367	15.5	58	0.58	129	3.7	26	33	48
Gibtsawit	8.7	35	24	25.7	546	1110	370	14.5	75	1.00	130	2.3	20	60	70
Wuahir	8.4	30	26	55.9	809	1673	1527	19.0	81	1.05	86	2.4	17	55	77
Zhara	8	29	22	50.8	286	1186	764	18.3	70	1.00	41	2.2	34	60	71
Asolit	7.7	36	24	4.8	267	1010	705	15.0	18	0.74	14	1.3	13	47	73
Mosha	7	43	32	57.8	1553	2421	5263	17.9	86	2.17	42	0.0	15	50	73
Gunaguna	6.9	34	24	24.3	457	638	229	19.2	81	1.19	138	1.1	22	52	72
Zagua	6.8	33	29	30.5	776	1659	1671	22.9	85	1.15	94	1.1	25	56	74
Enshet Kuskuam	6.3	26	19	18.5	452	990	476	10.1	55	0.69	52	1.2	24	61	83
Mantogera	5	44	29	30.5	844	1411	0	15.2	60	0.78	149	4.4	20	50	66
Mekedesemariam	4.7	33	21	80.2	700	1411	2533	12.6	78	1.67	33	0.9	33	59	77
Hiruy	4	31	26	111.5	620	2250	1850	20.0	85	2.41	29	0.2	22	58	78
Emashenkor	3.4	38	25	42.4	438	1194	1075	15.4	62	1.63	12.5	0.0	9	42	61
Quar	3	34	25	100.5	330	731	246	15.3	60	0.54	69	3.3	29	48	66
Ascha	1.6	22	18	28.9	900	1289	357	19.3	76	0.55	128	3.2	34	72	84

Table 3 Species richness measures and diversity indices for the 28 forests

Name of Church	Obser. Species	Chao1 Estim.	Jacknife1 Estim.	Rarefied* Spp no.	Diversity indices**		
					Sh-Wi	Sim	Fisher
Wonkesht	78	78	88	45	3.8	27.0	31.3
Gedamselase	74	99	100	40	3.6	20.6	26
Qorata	60	67	69	35	3.3	15.1	18.3
Shamo	58	67	75	44	3.8	41.5	28.6
Mantogera	44	52	59	31	3.1	13.6	15.4
Mosha	43	51	53	27	3.1	14.3	11.8
Gelawdios	42	47	48	28	3.1	14.8	12.2
Amstya	41	54	57	29	3.0	11.9	14.5
Woji	39	39	46	23	2.8	10.9	9.3
Emashenkor	38	44	52	31	3.1	14.8	14.2
Alember	37	50	49	20	2.5	6.5	9.8
Asolit	36	48	46	17	3.0	12.5	12.7
Dengolt	36	50	48	28	2.4	7.3	8.3
Gibtsawit	35	52	48	25	2.8	10.5	14
Gunaguna	34	49	48	23	3.0	15.8	14.4
Debresena	34	34	40	34	2.8	10.3	10
Quar	34	50	50	29	3.1	16.1	18.9
Zagua	33	57	44	24	2.8	10.6	9.7
Mekedesemariam	33	37	48	25	2.9	14.1	10.1
Hiruy	31	68	43	21	2.6	8.5	8.8
Wuahir	30	62	40	21	2.7	11.4	7.9
Zhara	29	46	44	24	2.5	7.0	9.8
Enshet Kuskuam	26	27	33	22	2.8	14.1	7.9
Dedim	22	22	26	17	1.9	3.9	5.5
Deder	22	34	31	13	1.6	3.5	5.2
Ascha	22	32	27	23	2.2	5.6	9.4
Abegeldi	19	22	22	14	1.9	4.0	4.5
Wonberoch	15	17	19	10	1.6	3.7	3

* Rarefied species number was calculated from common abundance level (98 which was the lowest total individual at Ascha from 28 forests)

** Sh-Wi-Shannon-Wiener's diversity index; Sim- Simpson diversity index; Fisher- Fisher alpha diversity index

Table 4. Regression analyses of selected dependent variables on independent variables altitude, forest area, cattle trails and dead stumps for 28 church forests. Summary of regression coefficients and p values for each independent factor against dependent variables. All bold cells show significant relations.

Independent variables	Coefficients	Dependent variables												
		Species richness							Structure				Similarity	
		Entire forest		Overstorey		Understorey		Ratio of richness *	Species Dominance			Ratio of Density**	Between under and overstorey	Between forests
		Observed (log)	Rarefied (log)	Observe (log)	Rarefied (log)	Observe (log)	Rarefied (log)		IVI1 (log)	IVI5 (log)	IVI10 (log)			
Altitude (km)	β_0	**2.2 P=0.000**	**1.96 P=0.000**	**1.95 P=0.000**	**1.5 P=0.000**	1.6 P=0.000	1.07 P=0.04	0.56 P=0.14	**0.54 P=0.04**	**1.3 P=0.000**	**1.5 P=0.000**	3.2 P=0.03	0.38 P=0.3	
	β_1	**-0.28 P=0.000**	**-0.24 P=0.001**	**-0.23 P=0.007**	**-0.15 P=0.04**	-0.32 P=0.05	-0.21 P=0.32	-0.06 P=0.69	**0.34 P=0.003**	**0.2 P=0.000**	**0.44 P=0.000**	-0.67 P=0.26	0.05 P=0.74	
	r^2	**0.40**	**0.34**	**0.26**	**0.15**	0.13	0.04	0.0	**0.30**	**0.40**	**0.45**	0.05	0.0	
Area (log ha)	β_0	1.4 P=0.000	1.4 P=0.000	1.29 P=0.000	1.24 P=0.000	0.69 P=0.000	0.67 P=0.004	0.35 P=0.02	1.4 P=0.000	1.7 P=0.000	1.87 P=0.000	1.7 P=0.008	**0.83 P=0.000**	
	β_1	0.08 P=0.28	0.03 P=0.6	0.03 P=0.5	0.06 P=0.16	0.07 P=0.26	0.04 P=0.63	0.02 P=0.67	-0.01 P=0.91	-0.03 P=0.62	-0.02 P=0.63	-0.05 P=0.8	**-0.31 P=0.02**	
	r^2	0.05	0.01	0.02	0.07	0.05	0.0	0.0	0.0	0.0	0.0	0.00	**0.2**	
Cattle trail/plot	β_0	1.6 P=0.000	1.5 P=0.000	1.4 P=0.000	1.2 P=0.000	**1.1 P=0.000**	**0.92 P=0.000**	0.5 P=0.000	**1.2 P=0.000**	1.67 P=0.000	1.8 P=0.000	**2.7 P=.000**	0.56 P=0.000	
	β_1	-0.04 P=0.11	-0.04 P=0.06	-0.02 P=0.54	-0.05 P=0.16	**-0.1 P=0.02**	**-0.16 P=0.004**	-0.05 P=0.2	**0.09 P=0.006**	0.03 P=0.1	0.01 P=0.35	**-0.51 P=0.001**	-0.03 P=0.5	
	r^2	0.09	0.1	0.01	0.07	**0.2**	**0.28**	0.06	**0.26**	0.1	0.03	**0.35**	0.02	
Dead stumps/ha	β_0	1.6 P=0.000	1.5 P=0.000	1.39 P=0.00	1.12 P=0.000	1.08 P=0.000	0.87 P=0.000	0.56 P=0.000	1.2 P=0.000	1.67 P=0.000	1.8 P=0.000	**3.6 P=.000**	0.58 P=0.000	
	β_1	-0.001 P=0.24	-0.001 P=0.17	0.0 P=0.77	0.0 P=0.7	-0.003 P=0.07	-0.004 P=0.08	0.0 P=0.15	0.0 P=0.07	0.0 P=0.1	0.0 P=0.32	**-0.04 P=0.01**	-0.001 P=0.43	
	β_2											**0.0002 P=0.05**		
	r^2	0.05	0.07	0.0	0.0	0.1	0.17	0.07	0.1	0.1	0.04	**0.3**	0.02	
Altitude*Cattle trail	β_0					**1.2 P=0.000**	**0.92 P=0.000**							
	β_1					**-0.05 P=0.003**	**-0.06 P=0.002**							
	r^2					**0.29**	**0.33**							
Altitude difference (km)	β_0													**0.48 P=0.000**
	β_1													**-0.37 P=0.000**
	r^2													**0.34**
Distance (km)	β_0													**0.48 P=0.000**
	β_1													**-0.002 P=0.000**
	r^2													**0.05**

* ratio of understorey to overstorey rarefied species richness and ** ratio of understorey to overstorey density

CHAPTER 3

SOIL SEED BANKS IN CHURCH FORESTS OF NORTHERN ETHIOPIA: Implications for the conservation of woody plants

Alemayehu Wassie and Demel Teketay

Published in: Flora (2006) 201:32–43

Abstract

Church forests are sanctuaries for different organisms, ranging from microbes to large animals, which have almost disappeared in most parts of northern Ethiopia. Despite the actual and potential significance of these forests, studies and documented information on their bio-physical features and socio-economic setting are either scanty or totally lacking. A study was, therefore, carried out in seven of these church forests, namely Hiruy, Zhara, Gibtsawit, Gelawdios, Dengolt, Debresena and Ascha with the objective of assessing the composition, densities and spatial heterogeneity of soil seed banks.

The number of viable seeds in the soil samples, from both germination and sieving, corresponded to a seed bank density down to 9 cm in the soil of 7594 seeds m^{-2} at Ascha, 2064 seeds m^{-2} at Debresena, 4208 seeds m^{-2} at Dengolt, 3158 seeds m^{-2} at Gelawdios, 2754 seeds m^{-2} at Gibtsawit, 2759 seeds m^{-2} at Hiruy, and 1909 seeds m^{-2} at Zahra. The total number of species recorded was 50, representing at least 22 plant families, in the seven church forests, of which 13 were recorded from Ascha, 29 from Debresena, 26 from Dengolt, 19 from Gibtsawit, 22 from Gelawdios, 20 from Hiruy and 21 from Zahra. There was a significant difference in seed densities and number of species recovered from the soil seed banks of the seven church forests. Ascha had a significantly higher seed bank density than all other church forests. Dengolt exhibited a significantly higher number of species compared with other church forests. Of the identified species recorded from the seven church forests, 83% were herbs, 11% trees and 6% shrubs. Our results revealed that church forests accumulate large quantities of persistent seeds of herbaceous species in the soil while only five (6%) of the 91 woody species recorded in the standing vegetation of the seven church forests were represented in the soil seed banks.

The fact that most of the dominant tree species do not accumulate seeds in the soil suggests that their regeneration from seeds would be prevented by removal of mature individuals in the standing vegetation. The results also provided further evidences that consolidate the conclusions of previous studies, i.e. the future existence of the woody flora characteristic of dry Afromontane areas in Ethiopia depends on the conservation and sustainable utilization of the few remnant natural forests.

1. Introduction

Dry Afromontane forests have disappeared from most of the highlands in northern Ethiopia, except around churches and some inaccessible pockets owing to deforestation associated with, mainly, expansion of farmlands, spurred by the rapidly increasing populations of humans and domestic animals, over centuries. Hence, a patch of indigenous old-aged trees in the northern highlands of Ethiopia signifies the presence of a church/monastery in the middle. These church forests, which are also referred to as 'debr' or 'geddam' locally, are visible from a great distance, with a majestic appearance, usually built on small hills overlooking the surrounding villages. They are sanctuaries for different organisms, ranging from microbes to large animals, which have almost disappeared in most parts of northern Ethiopia (Alemayehu Wassie, 2002; Alemayehu Wassie et al., 2005a&b).

The church forests are administered by the Ethiopian Orthodox Tewahido Church (EOTC), which is an indigenous and integral Christian Church of Africa, one of the oldest Churches in the world and founding member of the World Council of Churches (Aymro Wondmagegnehu and Motovu, 1970; Alemayehu Wassie, 2002). In addition to its religious activities, EOTC has also a long history of conservation of forest resources, which usually envelop the churches. Although the main purpose of churches is as places for worship, burials and meditating religious festivals, they also provide valuable, often unique, and secured habitats for plants, animals and microorganisms as well as green spaces for people to relieve the stressed mind. Church compounds are the safe places for trees and other biodiversity resources where one can animate trees that escaped from being destroyed forever under the shelter of the church value and esteem. In other words, church compounds are serving as in situ conservation and hot spot sites for biodiversity resources, mainly indigenous trees and shrubs of Ethiopia, which, in turn, give prestige for the religious sites (Alemayehu Wassie, 2002).

Despite the actual and potential significance of the church forests, studies and documented information on their biophysical and social dimensions are either scanty or totally lacking. Cognizant of this reality, a series of studies have been initiated focusing on the opportunities, constraints and prospects of EOTC in conserving forest resources (Alemayehu Wassie, 2002), diversity and regeneration status of woody plants and socio-economic setting associated with church forests (Alemayehu Wassie et al., 2005). The present study, which forms a component in the series of studies initiated, deals with the soil seed banks of church forests.

The soil seed bank refers to all viable seeds and fruits present on or in the soil and associated litter/humus. Soil seed banks can be either transient, with seeds that germinate within a year of initial dispersal, or persistent, with seeds that remain in the soil more than 1 year (Thompson and Grime, 1979; Leck et al., 1989; Simpson et al., 1989). They exhibit variations in space as well as time and display both horizontal and vertical dispersion, reflecting initial dispersal onto the soil and subsequent movement (Simpson et al., 1989). Soil seed banks reflect partly the history of the vegetation and can play an important role in its regeneration or restoration after disturbances. They have been exploited in two contexts: to manage the composition and structure of existing vegetation and to restore or establish native vegetation (Van der Valk and Pederson, 1989).

The pool of long-lived seeds in the soil accumulates over many decades and contributes as a source of propagules that ensure continual occupation of a site after disturbances while serving as a gene pool by buffering genetic changes in the population (Harper, 1977; Grime, 1979; Silvertown, 1982; Fenner, 1985; Hill and Morris, 1992). Soil seed banks play a crucial role in the dynamics of plant populations. In forest management, natural seed banks play a vital role in regeneration after disturbances, for example tree felling. The fact that many economically important trees are canopy species whose seeds have little dormancy makes it important to leave some individuals of the species to act as local seed sources. Knowledge of which species are not represented in the persistent soil seed bank can be just as important as knowing which species are represented. This is especially true in the management of vegetation for conservation. The dynamics of a soil seed bank include recruitment into the dormant seed bank population through seed rain, losses from the dormant seed bank through seed predation or death and transfer into the active seed bank to germinate and form a seedling bank, through a stimulus.

Apart from information on the occurrence of viable seeds in the soil of dry Afromontane forests and arable land (Demel Teketay and Granström, 1995; Demel Teketay, 1996, Demel Teketay, 1997a and Demel Teketay, 1998a), nothing is known about the mechanisms of incorporation of these seeds into the soil. In these areas, ground digging by bush pigs, trampling by grazing animals, scatter-hoarding rodents, ants, entrance of smaller seeds into cracks formed in the soil during the dry period and cultivation in arable land are among the factors responsible for the movement of seeds in the soil (own observation), although dung beetles (Estrada and Coates-Estrada, 1986), entrance of seeds in tunnels created by burrowing animals and decomposing roots or uprooted trees (Hopkins and Graham, 1983;

Putz, 1983; Young, 1985), earthworms (McRill and Sagar, 1973), termites (Lal, 1987; Garwood, 1989) rain washing seeds into the soil (Hopkins and Graham, 1983) and birds (Fenner, 1995) have been shown to be important.

The specific objective of this study was to assess the composition, density and spatial distribution of soil seed banks in seven church forests found in northern Ethiopia.

2. Materials and Methods

2.1. Study Sites

The study was conducted in South Gondar Administrative Zone (SGAZ), one of the Administrative Zones in the Amhara National Regional State, Northern Ethiopia. Its geographical location is between 11°02′–12°33′N and 37°25′–38°41′E with an altitude range of 1500–4231 m. SGAZ has an area of about 14,299 km^2. It has 10 Administrative Districts, known locally as 'Woredas' with a total population of 2,050,539 of which 95% of total population are members of the EOTC affiliated in 1404 churches (SGAZDPED unpublished). The annual average rainfall varies between 700 and 1300 mm and the annual average temperature ranges between 9.3 and 23.7 °C. The total forest cover in SGAZ is 20,882 ha, comprising 16,660 ha of natural forest and 4222 ha of man-made plantations (SGAZDA unpublished). The forest cover accounts for only 1.4% of the total area of SGAZ.

Seven churches were selected for the actual study from SGAZ, which were found surrounded by forests/trees ranging from 1.6 to 100 ha (Table 1 and Table 2). The churches selected were located at different altitude, ranging from 1900 to 3100 m, and established in different periods dating from year 368 to 1948. From top view position, the churchyards are almost circular in shape. The churches are immediately surrounded by open space for worship and festivals, which in turn is enveloped by the forest.

The dry Afromontane forests (Demel Teketay, 1996), to which the studied church forests belong, are either *Juniperus-Podocarpus* forests or predominantly *Podocarpus* forests, both with an element of broad-leaved species. They occur at altitudes from 1500 to 2700 m, with an average annual temperature between 14 and 20 °C and annual rainfall between 700 and 1100 mm, with most of the rain in July (Friis, 1992). The diversity and regeneration status of woody plants in the study forests has been reported elsewhere (Alemayehu Wassie et al., 2005a).

Table 1. The seven church forests with their areas, number of species and plant families, total density, basal area, number of live stems (dbh ≥ 10 cm) and dead stumps

Church	**Time/year of establishment**	**Species**	**Families**	**Density[a]**	**Basal area[b]**	**Live stems[a]**	**Dead stumps[a]**
Gelawdios	1508 AD	42	29	1109	52	352	32
Dengolt	1305 AD	36	28	1169	35	268	41
Gibtsawit	1628 AD	35	24	1110	25	490	130
Debresena	First half of the 16th	34	27	2141	45	1003	119
Hiruy	368 AD	31	26	2250	110	507	29
Zahra	Mid of 16th century	29	22	1186	50	227	41
Ascha	1948 AD	22	18	1289	28	786	128

Source: Alemayehu Wassie et al. (2005).

[a] Individuals ha^{-1}; [b] $m^2\ ha^{-1}$

2.2. Data collection

In all the church forests, parallel transects were laid out. The first transect was aligned randomly at one side of the forest using a compass; then the others were laid systematically at 50 m intervals from each other. Then, along the first transect line a 10 m×10 m quadrat was marked randomly and then the same size of the plots were marked at 50 m intervals systematically on all other transects. Since the forest at Gelawdios was so extensive, the transects and sample plots were laid at 100 m intervals.

In each of the quadrats, five plots measuring 15 cm×15 cm were laid, one at the center and the other four at the four corners. From each plot, three soil layers of 3-cm thickness each, i.e. a total depth of nine centimeters, were collected using a knife and spoon. Then, similar layers from these five plots within a quadrat were mixed to form a soil composite in order to reduce variability within the quadrats. The composite sample for each soil layer was again divided into five equal parts among which one was selected randomly as a working soil sample (Tefera Mengistu, 2001; Emiru Birhane, 2002; Tefera Mengistu et al., 2005). A total of 570 soil samples from 190 quadrats (Hiruy-14; Zahara-22; Gibtsawit-19; Gelawdiouose-44; Dengolt-52; Debresena-32; and Ascha-7) were collected. On the collection days, the soil samples were transported from the sites of origin to a safe storing place. Then, the soil samples were transported to the Forestry Research Center (FRC) where they were sieved to recover seeds of woody plants. Four sieve sizes, i.e. 1, 1.6, 2 and 3.15 mm, were selected assuming that seed sizes of the different species are with in these ranges. Viability of seeds recovered by sieving was determined by cutting tests (Demel Teketay and Granström, 1995) after they were identified.

The sieved soils were, then, transported to the Ethiopian Agricultural Research Organization (EARO) headquarters where they were incubated in a glasshouse to stimulate germination of seeds. In the glasshouse, the soil samples were spread as thinly as possible on plastic trays and watered every day. Seedlings started germinating from the soil samples within a week. The seedlings were identified, counted and removed. Those that were difficult to identify were transplanted into polyethylene bags filled with a soil medium and left to grow until they lend themselves for identification. Those difficult to identify were categorized as unidentified species. Thus, soil sieving together with seedling emergence method was used to assess the status of the soil seed banks.

2.3. Data analyses

To compare similarities of species composition of the soil seed banks between the church forests as well as between soil seed bank and standing vegetation of woody species in each church forest, Jaccard's Coefficient of Similarity (S_j) was employed (Krebs, 1989). The data on species numbers and seed densities from the seven church forests were subjected to one-way ANOVA (Zar, 1984). Following the ANOVA, Tukey's HSD multiple comparison was used to test for significant differences among the means.

Plant nomenclature

Plant nomenclature in this paper follows those of Cufodontis (1953–1972), Friis (1992), Hedberg and Edwards (1989), Edwards et al. (1995), Hedberg and Edwards (1995), Edwards et al. (1997), Edwards et al. (2000) , and Hedberg et al. (2003).

3. Results

Species composition and similarities

There was a significant difference in the number of species recovered from the soil seed banks of the seven church forests [One-Way ANOVA: $F_{(6,183)}=9.28$, $P<0.0001$]. Dengolt exhibited a significantly higher ($P<0.05$) number of species compared with other church forests except Debresena (Table 2). The total number of species recorded was 50, representing at least 22 plant families, in the seven church forests, of which 13 were recorded from Ascha, 29 from Debresena, 26 from Dengolt, 19 from Gibtsawit, 22 from Gelawdios, 20 from Hiruy and 21 from Zahra (Table 3). Of the identified species recorded from the seven church forests, 86% were herbs and 14% woody plants. Only five of the species occurred in all of the seven church forests while three occurred in six of them.

The similarity in species composition of the soil seed bank between the church forests was generally low and ranged from S_j values of 0.26 (between Ascha and Gelawdios as well as Ascha and Zahra) to 0.53 (between Debresena and Dengolt). The second highest similarity in species composition (S_j=0.50) was recorded between Dengolt and Gibtsawit (Table 4). The similarity between the soil seed bank and standing vegetation of woody species was very low (ranging from S_j values of 0.02 for Dengolt to 0.14 for Gibtsawit).

Densities of seeds in the soil

Densities of seeds in the soil collected from the seven church forests also showed significant differences [One-Way ANOVA: $F_{(6,183)}=12.68$, $P<0.0001$]. Ascha had a significantly higher ($P<0.05$) seed bank density than all other church forests (Table 2). The number of viable seeds in the soil samples, from both germination and sieving, corresponded to a seed bank density down to 9 cm in the soil of 7594 seeds m^{-2} at Ascha, 2064 seeds m^{-2} at Debresena, 4208 seeds m^{-2} at Dengolt, 3158 seeds m^{-2} at Gelawdios, 2754 seeds m^{-2} at Gibtsawit, 2759 seeds m^{-2} at Hiruy and 1909 seeds m^{-2} at Zahra (Table 2).

The three species with the highest soil seed densities included, in descending density, *Juniperus procera, Trifolium* sp. and *Poa leptoclada* at Ascha, *Impatiens hochstetteri, Trifolium* sp. and *P. leptoclada* at Debresena, *Trifolium* sp., *J. procera* and *P. leptoclada* at Dengolt, *Trifolium* sp., *P. leptoclada* and *I. hochstetteri* at Gelawdios, *Galinsoga parviflora, Trifolium* sp. and *P. leptoclada* at

Gibtsawit, *P. leptoclada, Trifolium* sp. and *Dicrocephala integrifolia* at Hiruy and *Trifolium* sp., *P. leptoclada* and *Sida rhombifolia* at Zahra (Table 4).

The most frequent species (frequency=90–100%) in the soil seed banks were *Laggera crispata* and *Trifolium* sp. (each with frequency values of 100%) at Ascha, *Dicrocephala integrifolia, Geranium simense, P. leptoclada* and *Trifolium* sp. at Dengolt, *P. leptoclada* and *Trifolium* sp. (each with frequency values of 100%) at Gelawdios, *G. parviflora* (frequency=100%) at Gibtsawit, *Dicrocephala integrifolia* (frequency=100%), *G. simense, L. crispata* and *P. leptoclada* at Hiruy and *L. crispata* and *Trifolium* sp. (frequency=100%) at Zahra (Table 4). At Debresena, the two most frequent species were *Dicrocephala integrifolia* (frequency=84%) and *Trifolium* sp. (frequency=78%).

Depth and horizontal distribution of seeds

The depth distribution of the whole seed bank was more or less consistent at all church forests, except Hiruy and Zahra, with the highest densities in the upper three centimeters of soil and then gradually decreasing densities with increasing depth (Fig. 1). The number of species present showed a similar trend (Fig. 2). There was more variation among species than sites in relation to depth distribution (Table 5).

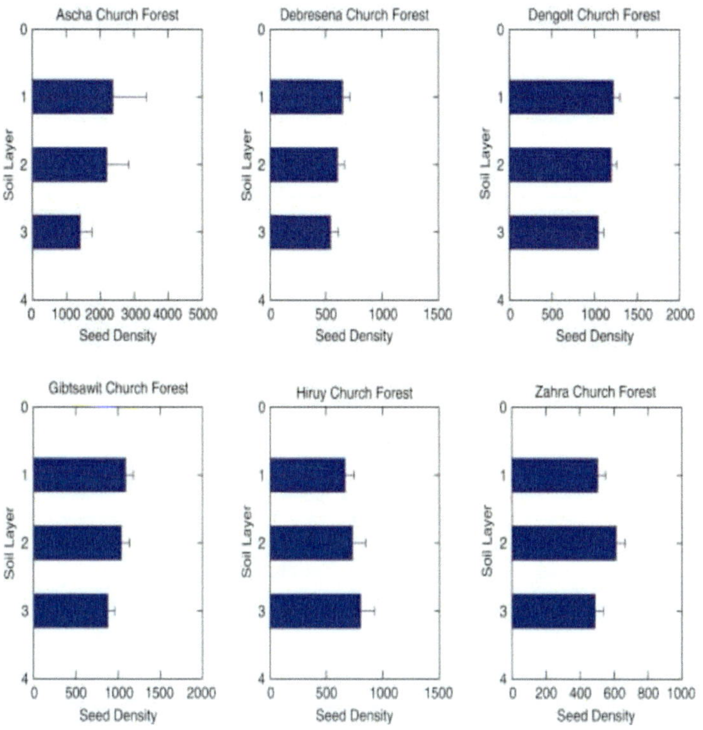

Figure 1. Depth distribution of seeds recovered from soil samples collected in the seven church forests.

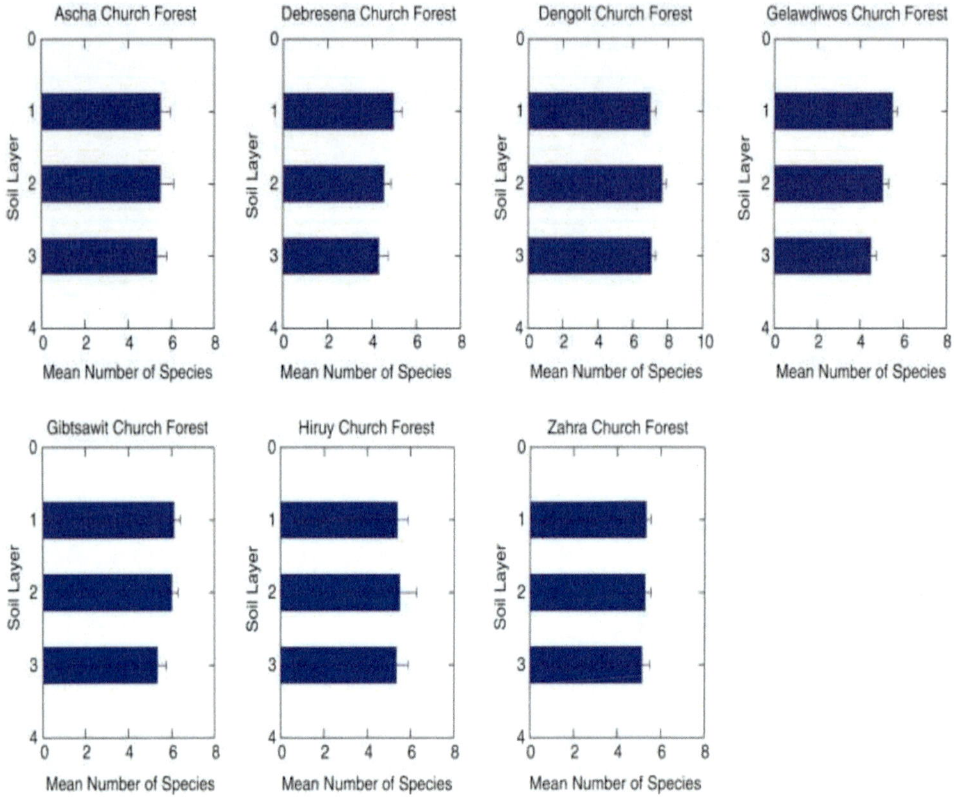

Figure 2. Depth distribution of species recovered from soil samples collected in the seven church forests.

The horizontal distribution of species and seeds in the soil exhibited considerable variation as can be verified from the wide ranges of number of species and soil seed densities recovered from soil sample plots in each of the church forests (Table 2; Fig. 3).

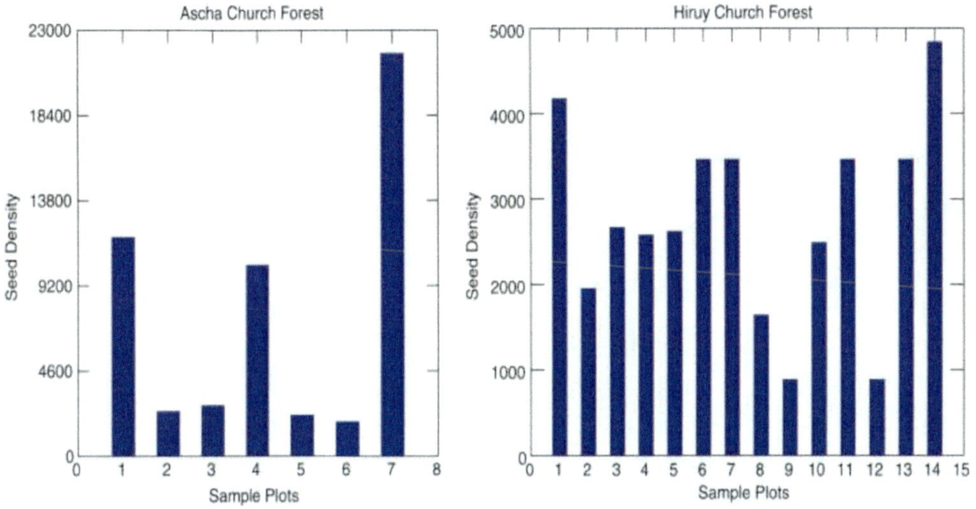

Fig. 3. Horizontal heterogeneity of seeds recovered from soil samples collected in Ascha and Hiruy church forests.

Soil sieving

The total number of seeds recovered through sieving of soil samples collected in the seven church forests was 2470 representing eight identified species, namely *Calpurnia aurea, Chionanthus mildbraedii, Cordia africana, Croton macrostachyus, Ekebergia capensis, Diospyros* sp., *J. procera* and *Olea europaea* subsp. *cuspidata* (Table 6). None of these species germinated from the soil samples incubated in the glasshouse. The majority of seeds (about 68%) belonged to *J. procera,* with a relatively higher number of seeds at Ascha and Dengolt. The share of *Chionanthus mildbraedii* and *O. europaea* seeds was about 21% and 9%, respectively, while proportions of seeds of the other five species were one or less than one percent. Of the total seeds obtained from sieving, 1559 (63%) were viable while 911 (37%) were dead. More than 98% of all the viable seeds recovered belonged to *J. procera*. While all seeds of *Calpurnia aurea* were viable, 92%, 50%, 18% and 7% of seeds of *J. procera, Diospyros* sp., *Croton macrostachyus* and *Cordia africana,* respectively, were viable and none of the seeds of *Chionanthus mildbraedii, E. capensis* and *O. europaea* were viable. Although intact seeds of

Podocarpus falcatus were not recovered, woody seed coats broken into halves, which indicate germination of the seeds, were seen during sieving.

Table 2. Mean number of species (± SE) and seed densities (± SE) recovered from soil samples collected in the seven church forests.

Church forest	Area of forest (ha)	Number of samples	Mean number of species	Range in the number of species	Mean density of seeds	Range in the densities of seeds
Ascha	1.6	7	9 ± 0.9[a]	6 - 19	7594 ± 2832[a]	1911 - 9911
Debresena	11.5	32	10 ± 0.6[ab]	5 - 16	2064 ± 230[e]	711 - 7067
Dengolt	25	52	12 ± 0.4[b]	6 - 19	4208 ± 241[c]	1911 - 9200
Gelawdios	100	44	9 ± 0.4[a]	6 - 14	3158 ± 298[ce]	667 - 7067
Gibtsawit	8.7	19	9 ± 0.3[a]	7 - 13	2754 ± 178[e]	1600 - 7067
Hiruy	4	14	9 ± 0.5[a]	5 - 12	2759 ± 309[ce]	889 - 4844
Zahra	8	22	9 ± 0.3[a]	7 - 14	1909 ± 183[e]	933 - 4489

Figures with different letter in a column are significantly different ($P < 0.05$) from each other.

Table 3. Density (D) and frequency (F) of species recovered from soil samples collected in the seven church forests.

SPECIES	FAMILY	ASC		DEB		DEN		GEL		GIB		HIR		ZAH	
		D	F	D	F	D	F	D	F	D	F	D	F	D	F
Achyranthus aspera	Amaranthaceae	-	-	24	19	-	-	-	-	363	68	-	-	-	-
Ageratum conyzoides	Asteraceae	248	43	-	-	-	-	-	-	-	-	10	21	-	-
Alchemilla cryptantha	Rosaceae	-	-	78	34	89	25	-	-	-	-	79	14	-	-
Arthraxon micans	Poaceae	114	71	-	-	-	-	-	-	-	-	-	-	-	-
Bidens pilosa	Asteraceae	133	71	51	34	49	27	31	27	65	32	51	29	61	27
*Calpurnia aurea**	Fabaceae	-	-	10	3	-	-	15	14	5	11	-	-	-	-
Caylusea abyssinica	*Resedaceae*	-	-	10	13	-	-	-	-	-	-	-	-	14	23
*Cordia africana**	*Boraginaceae*	-	-	-	-	-	-	-	-	-	-	-	-	4	9
Chenopodium murale	Chenopodiaceae	-	-	10	10	-	-	-	-	-	-	3	7	-	-
Chloris pycnothrix	*Fabaceae*	-	-	-	-	-	-	23	27	-	-	-	-	-	-
Crassula alsinoides	*Crassulaceae*	-	-	8	9	79	50	11	11	-	-	-	-	-	-
*Croton macrostachyus**	Euphorbiaceae	-	-	-	-	-	-	-	-	-	-	3	7	8	9
Cyperus rotundus	Cyperaceae	-	-	39	38	148	71	-	-	159	68	-	-	-	-
Dichrocephala integrifolia	Asteraceae	-	-	146	84	285	94	139	64	175	58	410	100	196	68
Digitaria velutina	Poaceae	-	-	14	13	-	-	22	16	-	-	-	-	-	-
Diospyros sp.*	Ebenaceae	-	-	-	-	-	-	-	-	-	-	-	-	4	5
Eleusine indica	Poaceae	-	-	-	-	9	8	2	5	112	47	32	36	30	32
Eragrostis tef	Poaceae	-	-	-	-	34	21	-	-	23	21	-	-	-	-
Galinsoga parviflora	Asteraceae	-	-	-	-	74	44	149	64	533	100	248	86	186	77
Geranium simense	Geraniaceae	229	86	172	63	432	94	146	66	14	16	337	93	127	86
Impatiens hochstetteri	Balsaminaceae	-	-	367	66	-	-	286	82	-	-	-	-	121	64
*Juniperus procera**	Cupressaceae	4927	71	7	9	616	65	-	-	58	11	-	-	-	-
Kalanchoe petitiana	Crassulaceae	-	-	13	19	36	38	22	20	-	-	19	7	-	-
Laggera crispata	Asteraceae	127	100	106	63	97	44	74	57	168	74	273	93	166	95
Misopates orontium	Scrophulariaceae	267	71	60	38	44	42	49	32	33	26	-	-	12	18
Oldenlandia monanthos	Rubiaceae	13	14	18	19	27	21	-	-	-	-	-	-	-	-
Persicaria nepalense	Polygonaceae	-	-	108	59	121	50	23	80	306	89	10	14	30	41
Phyllantus leucanthus	Euphorbiaceae	-	-	-	-	8	12	-	-	-	-	-	-	-	-
Physalis micrantha	Solanaceae	-	-	24	28	38	33	-	-	253	79	-	-	48	41
Pilea tetraphylla	Urticaceae	-	-	86	59	95	35	-	-	-	-	114	71	-	-
Poa leptoclada	Poaceae	419	86	182	69	587	96	636	98	395	89	467	93	309	86
Sida rhombifolia	Malvaceae	51	57	46	44	16	10	-	-	2	5	-	-	200	86
Trifolium sp.	Fabaceae	705	100	321	78	654	98	642	95	473	84	429	86	366	100
Veronica abyssinica	Scrophulariaceae	337	86	69	44	300	81	149	61	-	-	206	86	-	-
Veronica javanica	*Scrophulariaceae*	-	-	76	59	230	56	95	43	-	-	25	7	-	-
Vicia hirsuta	*Fabaceae*	-	-	8	9	-	-	-	-	14	21	10	14	6	9
Zehneria scabra	Cucurbitaceae	-	-	11	13	-	-	-	-	-	-	-	-	-	-
Unidentified species	-	25	-	29	-	51	-	23	-	5	-	35	-	10	-
Number of species	-	13	-	29	-	26	-	22	-	19	-	20	-	21	-

ASC = Ascha; DEB = Debresena; DEN = Dengolt; GEL = Gelawdios; GIB = Gibtsawit; HIR = Hiruy; and ZAH = Zahra; * = woody species

Table 4. Similarities in species composition of soil seed banks between the church forests.

	Ascha	Debresena	Dengolt	Gibtsawit	Gelawdios	Hiruy	Zahra
Ascha	-	0.31	0.38	0.33	0.26	0.27	0.26
Debresena		-	0.53	0.39	0.42	0.43	0.35
Dengolt			-	0.50	0.41	0.44	0.32
Gibtsawit				-	0.37	0.33	0.48
Gelawdios					-	0.45	0.37
Hiruy						-	0.37
Zahra							-

Table 5. Depth distribution of seeds of species recorded in more than four of the church forests.

Church forest	Layer	Density of seeds m^{-2}											
		BP	DI	EI	GP	GS	LC	MO	PN	PL	SR	TS	VA
Ascha	1	51	0	0	0	89	13	114	0	133	32	235	108
	2	63	0	0	0	44	76	51	0	133	6	235	102
	3	19	0	0	0	95	38	102	0	152	13	235	127
	Total	133	0	0	0	228	127	267	0	418	51	705	337
Debresena	1	19	53	0	0	51	29	14	57	92	11	92	18
	2	15	56	0	0	60	54	28	25	54	24	93	35
	3	17	38	0	0	61	22	18	26	36	9.7	136	17
	Total	51	146	0	0	172	106	60	108	182	44	321	69
Dengolt	1	21	103	3	29	174	16	11	38	233	8	255	142
	2	15	91	2	21	144	49	19	48	190	6	227	79
	3	13	91	2	25	114	32	15	34	164	3	172	79
	Total	49	285	7	74	432	97	44	121	587	17	654	300
Gelawdios	1	11	40	2	81	45	39	24	86	306	0	194	65
	2	11	56	0	37	53	20	10	73	215	0	231	57
	3	9.1	43	0	31	48	14	15	77	210	0	217	28
	Total	31	139	2	149	146	74	49	235	731	0	642	149
Gibtsawit	1	40	56	26	199	7	47	7	119	154	2	171	0
	2	23	68	49	171	2	49	19	77	138	0	175	0
	3	2.3	51	47	164	5	73	7	110	112	0	126	0
	Total	65	175	122	533	14	168	33	306	405	2	473	0
Hiruy	1	22	152	0	63	76	124	0	3	133	0	70	73
	2	19	111	19	67	149	92	0	7	156	0	194	54
	3	10	146	13	117	111	57	0	0	178	0	165	79
	Total	51	410	32	248	337	273	0	10	467	0	429	206
Zahra	1	16	48	14	55	26	69	8	8	67	67	99	0
	2	26	87	8	51	77	48	0	10	113	65	149	0
	3	18	61	8	81	26	48	4	12	127	69	117	0
	Total	61	196	30	186	129	166	12	30	307	200	366	0

BP = *Bidens pilosa;* DI = *Dichrocephala integrifolia*; EI= *Elusine indica*; GP = *Galinsoga parviflora*; GS = *Geranium simense*; LC = *Laggera crispata*; MO = *Misopates orontium*; PN = *Persicaria nepalense*; PL = *Poa leptoclada*; SR = *Sida rhombifolia*; TS = *Trifolium* sp.; and VA = *Veronica abyssinica.*

Table 6. Total number of seeds of identified woody species recovered by sieving the soil samples collected from the seven church forests.

Church Forest	Status of Seeds	Species								Total
		CA	*CM*	*CAf*	*CMa*	*DS*	*EC*	*JP*	*OE*	
Ascha	Viable	0	0	0	0	0	0	856	0	856
	Dead	0	0	0	0	0	0	0	0	0
	Total	0	0	0	0	0	0	856	0	856
Debresena	Viable	5	0	0	0	0	0	5	0	10
	Dead	0	0	0	7	0	0	0	0	7
	Total	5	0	0	7	0	0	5	0	17
Dengolt	Viable	0	0	0	0	0	0	642	0	642
	Dead	0	0	0	0	0	0	122	0	122
	Total	0	0	0	0	0	0	764	0	764
Gelawdios	Viable	15	0	0	0	0	0	0	0	15
	Dead	0	507	0	1	0	0	0	0	508
	Total	15	507	0	1	0	0	0	0	523
Gibtsawit	Viable	2	0	0	0	0	0	25	0	27
	Dead	0	0	4	0	0	2	18	209	233
	Total	2	0	4	0	0	2	43	209	260
Hiruy	Viable	0	0	0	1	0	0	0	0	1
	Dead	0	0	0	1	0	1	1	0	3
	Total	0	0	0	2	0	1	1	0	4
Zahra	Viable	0	0	2	4	2	0	0	0	8
	Dead	0	0	22	14	2	0	0	0	38
	Total	0	0	24	18	4	0	0	0	46
Total	Viable	22	0	2	5	2	0	1528	0	1559
	Dead	0	507	26	23	2	3	141	209	911
	Total	22	507	28	28	4	3	1669	209	2470

CA = *Calpurnia aurea;* CM = *Chionanthus mildbraedii**; CAf = *Cordia africana;* CM = *Croton macrostachyu;* DS = *Diospyros* sp.; EC = *Ekebergia capensis***; JP = *Juniperus procera;* OE = *Olea europaea**
Family: * = Oleaceae; ** = Meliaceae

4. Discussion

Our results revealed that church forests accumulate large quantities of persistent seeds of herbaceous species in the soil. The accumulation of seeds in the soil is favored by the dormancy of many of the seeds, which is caused by either the presence of embryo dormancy or impermeable seed coat or both. Dormancy is selected for in most of dry Afromontane species, which are characterized by long dry seasons and unreliable rainy periods. In non-dormant seeds, dormancy will be induced if the seeds are dispersed under dense canopy or buried in the soil (Demel Teketay and Granström, 1995;Demel Teketay, 1996).

Most of the woody component of the church forest vegetation lacked reserves of long-lived seeds in the soil. Of the 91 woody species recorded in the standing vegetation of the seven church forests (Alemayehu Wassie, 2002), only five species (6%) were represented in the soil seed banks. Our findings concur with those of different authors (Granström, 1982; Demel Teketay and Granström, 1995;Demel Teketay, 1996, Demel Teketay, 1997a and Demel Teketay, 1998a; Kebrom Tekle and Tesfaye Bekele, 2000; Feyera Senbeta and Demel Teketay, 2001; Feyera Senbeta and Demel Teketay, 2002;Feyera Senbeta et al., 2002; De Villiers et al., 2003; Ericksson et al., 2003; Amiaud and Touzard, 2004; Tefera Mengistu et al., 2005) who reported higher and lower proportions of herbaceous and woody species in the soil seed banks, respectively.

Seeds of several woody species are large and contain high moisture, indicative of adaptations to immediate germination and seedling establishment and survival under the canopy of forests (Demel Teketay and Granström, 1995;Demel Teketay, 1996 and Demel Teketay, 1997b). The residence time of their seeds in the soil is relatively short compared with those of herbaceous species, which are capable of maintaining their viability for a long time (Demel Teketay and Granström, 1997a&b; Demel Teketay, 1998b). Those seeds, which do not germinate are consumed by predators or succumb to attack by microorganisms. By immediate germination and establishment, many trees and shrubs form large populations of seedlings in the forest (Demel Teketay, 1997b). Many of them are also capable of re-sprouting from damaged stems or roots (Demel Teketay, 1997a). This implies that the sources of re-growth of woody species are totally dependent on the presence of the whole or a portion of the forest vegetation with mature individuals. In the event of disturbance, herbaceous species can regenerate from both the soil seed bank or from recently dispersed seeds while the woody species from pre-existing seedlings, coppice shoots or from recently dispersed seeds.

The highest number of species in the soil seed bank was recorded at Dengolt, which could be attributed to the fact that the forest was more intensely disturbed and encroached than the other church forests (Alemayehu Wassie, 2002) and, hence, it might have experienced a high turnover, which contributed to the higher number of herbaceous species in soil seed bank. On the other hand, the higher density of seeds found in Ascha may be attributed to the fact that Ascha is located at a higher altitude (above 3100 m), the deterioration of seeds could be lower due to lower temperature. Conversely, Zhara, which is located at 1900 m and with higher microbial and pathogenic infestation, exhibited the lowest seed density. The church forests exhibited low similarities in the species composition of their soil seed banks, which suggests that each church forest has its own unique species composition and vegetation history. Therefore, the studied church forests are equally important and sensitive from a conservation point of view.

The spatial distribution, both vertically and horizontally, of seeds of different species, sample plots in each church forest and the different church forests varied greatly. These variations may reflect differences of species in terms of seed longevity in the soil, mode of seed dispersal, seed predation and probably differences in altitude and local edaphic conditions where seeds land. This result is also in agreement with past similar works (Demel Teketay and Granström, 1995; Demel Teketay, 1996; Demel Teketay, 1997a; Demel Teketay, 1998a; Feyera Senbeta and Demel Teketay, 2001; Feyera Senbeta and Demel Teketay, 2002; Feyera Senbeta et al., 2002; Ericksson et al., 2003; Matus et al., 2005; Tefera Mengistu et al., 2005).

Seed bank species have been reported to have the following characters related to their success in establishing after disturbance: production of numerous small seeds, means of long distance dispersal, formation of persistent soil seed banks and the capacity to remain viable in a dormant state for a long period of time (Thompson, 1987; Demel Teketay, 1998a,b,c,d &e ;Demel Teketay, 2002). Previous germination tests with some of the seed bank forming species in Ethiopia (Demel Teketay and Granström, 1995; Demel Teketay and Granström, 1997a&b; Demel Teketay, 1998b&c) revealed that seed dormancy in the species are controlled by requirements for mechanical scarification, e.g. *Calpurnia aurea*; light, e.g. *Croton macrostachyus* and *J. procera;* both light/light with relatively high red:far red ratio (no germination in darkness) and alternating temperatures for germination in fresh seeds, e.g. *Alchemilla cryptantha, Crassula alsinoides, Digitaria velutina, Veronica abyssinica* and *Veronica javanica*; and only light/light with relatively high red:far red ratio (no germination in darkness), e.g. *Dichrocephala integrifolia, G. parviflora, L. crispata* and *P. leptoclada.* Since seeds of most of these species are deeply buried, dormancy of the seeds can be broken only in connection with disturbances.

The higher proportion of *J. procera* seeds from the soil sieving and also their higher viability may indicate that this species possesses a higher potential for forming soil seed banks, which could ensure its perpetuation in the events of disturbances. Densities of seeds were higher at Ascha and Dengolt where *J. procera* was found with higher frequency (85% and 70%, respectively) in the standing vegetation (Alemayehu Wassie, 2002; Alemayehu Wassie et al., 2005a), indicating that the standing vegetation has significant contributions to its soil seed bank. However, during the study of the standing vegetation (Alemayehu Wassie, 2002; Alemayehu Wassie et al., 2005a), only a few seedlings of *J. procera* were observed, which might imply constraints, i.e. unfavorable environmental conditions for germination and recruitment or seedling mortality in these two church forests. The species has also been shown to accumulate in soil seed banks of other dry Afromontane forests of Ethiopia (Demel Teketay and Granström, 1995). Although the share of *Chionanthus mildbraedii* in the seeds recovered from sieving of the soil was the next highest amount, all of the seeds were not viable. The fact that the species had 57% frequency in the standing vegetation of Gelawdios may indicate that these seeds have a short viability and, hence, prefer to germinate immediately and form seedling banks.

Dry Afromontane church forests can be characterized by possessing large populations of buried seeds of herbs, grasses and sedges whereas trees mainly have persistent seedling banks and the ability to sprout from damaged roots or shoots. The fact that most of the dominant tree species do not accumulate seeds in the soil suggests that their regeneration from seeds would be prevented by removal of mature individuals in the standing vegetation. This, in turn, implies that restoration of church forests would be difficult and slow to accomplish if they are destroyed. On the other hand, the herbaceous vegetation has a better chance of recovery since it has a diverse soil seed bank with great seed longevity. Our results provide further evidences that consolidate the conclusions of previous studies on soil seed banks, i.e. the future existence of the woody flora characteristic of dry Afromontane areas in Ethiopia depends on the conservation and sustainable utilization of the few remnant natural forests.

CHAPTER 4

POST-DISPERSAL SEED PREDATION AND SEED VIABILITY IN FOREST SOILS: implications for the regeneration of tree species in Ethiopian church forests

Alemayehu Wassie, Tesfaye Bekele, Frank Sterck, Demel Teketay and Frans Bongers

Abstract

Dry Afromontane forests of Ethiopia have faced vast exploitation. Almost all forests have been converted to agricultural lands, except for small fragments that are left in the most inaccessible areas or around churches ("Church forests"). We investigated post-dispersal seed predation and the long term behavior of seeds in forest soil of selected tree species from these forests, and their implications for regeneration of trees in church forests.

We addressed the questions (1) How intense is post-dispersal seed predation in church forest, and does this seed predation vary with species and/or habitat, and (2) for how long tree seeds maintain their viability while buried in forest soil. Each question is addressed in one experiment. In the seed predation experiment we monitored seeds of six tree species in four habitats for a period of 14 weeks. In the seed viability experiment we assessed seed viability of five species in four sites after being buried 6, 12, or 18 months.

92% of the tree seeds was predated within 3.5 months. Predation was mainly dependent on species whereas habitat had a weaker effect. Predation was higher in closed canopy plots than in gaps. Plots in forest interior and edges were not different. Over time the differences in seed survival between species and between gaps and closed canopy increased.

Seed viability decreased sharply with burial time in soil for all species except for seeds of *Juniperus*, which still had 91% of viability after 18 months. Species significantly affect the viability of the seeds after 18 months of burial. The quick decline of seed viability in forest soil indicates that the study species do not have a persistent soil seed bank.

An absent or only transient soil seed bank of the studied species, combined with a high seed predation suggests a seed availability limitation for regeneration in the forest. In order to maintain such species in the forest the standing vegetation needs to be persistently managed and conserved. A continuous seed rain coming from a large enough population of adult trees may safeguard successful regeneration. Additional seed sowing, and seed and seedling protection (by e.g. animal enclosures) may increase successful regeneration of important species in these forests.

1. Introduction

Tropical dry forests are among the most exploited forests and have been converted to small fragments since long (Cabin et al., 2002). Habitat fragmentation has resulted in both a reduced area and an increased isolation of the remaining forest patches (Saunders et al., 1991; Fleury and Galetti, 2004). Regeneration of many of the tree species that inhabit these fragmented forests is becoming a critical problem, and further limitation may lead to the entire elimination of some of the species. Dry Afromontane forests of Ethiopia have been exploited to the extent that almost all these forests have been converted into agricultural lands (EFAP, 1994; Demel, 1996; Tesfaye et al., 2003), except for small fragments that are left in the most inaccessible areas or around churches ("Church forests") (Bingelli et al., 2003; Alemayehu Wassie et al., 2005 a&b; Aerts et al., 2006a).

An important indirect fragmentation effect is the change in viable seed availability. Lower production of seeds, changed seed dispersal and increased seed predation may hamper successful recruitment and establishment of trees and may affect vegetation dynamics (De Steven 1991; Heske et al. 1993; Hoffmann et al. 1995; Wijdeven and Kuzee 2000; Kollmann and Buschor, 2002; Donoso et al, 2003). Once tree seeds reach the forest floor, their fate could be immediate germination (and formation of seedling banks), predation, microbial degradation or burial (and formation of soil seed banks), depending on species and local environment (Whitmore, 1991; Vazquez-Yanes and Orozco-Segovia, 1993; Demel Teketay and Granström, 1997). Seeds represent a particularly concentrated source of potential food of high nutritive quality to any organism able to exploit it. Therefore, it is not surprising to find that in many plant species, a large proportion of seed production is lost to predation (Fenner and Kitajima, 1999).

Predation reduces the chances for reproduction. Especially when regeneration is limited by seed numbers, the consequence of seed predation for plant regeneration and population growth is very significant (Crawley, 1992; Hau, 1997; Fenner and Kitajima, 1999). It may also influence the genetic makeup of the plant population by differential selection of seeds even when it has little effect on recruitment (Benkman, 1995; Fenner and Kitajima, 1999). Two distinct groups of predation exist: pre-dispersal and post-dispersal predation (Fenner and Kitajima, 1999). Since post-dispersal seed removal may ultimately determine the fate of seeds, it might have a major impact on individual reproductive success of plants, spatial distribution and community structure of many plant species (Janzen, 1970; Connell, 1971; Howe & Smallwood 1982; Price & Jenkins 1986; Hulme 1996;

Sánchez-Cordero and Martínez-Gallardo, 1998; Hulme and Borelli, 1999; Fenner and Kitajima, 1999).

Several studies have shown that seed predation may vary with seed species, type of predators, seed size, density and burial depth as well as habitat (gaps, closed canopy, forest edge or interior) (Janzen 1971; Myster and Pickett, 1993; Sánchez-Cordero and Martínez-Gallardo, 1998; Hulme and Borelli, 1999; Kollmann and Buschor, 2002). Ethiopian church forests, in most cases surrounded by a farm dominated landscape, are preferred homes for many mammals, rodents, birds and insects. Since much of the surrounding farmed landscape is unsuitable for the survival of these animals, they tend to migrate to these remnant forests. It is obvious that fruits and seeds are major sources of food for the animals in these forests. As a result, seed predation may play a crucial role in the regeneration of tree species in church forest.

If seeds on the forest floor do not germinate immediately and escape from predation and microbial degradation, they enter into the soil seed bank. In general, soil seed bank forming species have seeds with an environmental sensor that delays germination until specific conditions of light and temperature are met (Bazzaz and Pickett, 1980; Vazquez-Yanes and Orozco-Segovia, 1993; Demel Teketay and Granström, 1997). For how long seeds can stay buried in the soil while maintaining viability depends mainly on the type of species (Garwood 1989; Thompson 1992; Fenner 1995; Demel Teketay and Granström, 1997).

We studied the effect of post-dispersal seed predation and the long-term behavior of seeds in forest soils of indigenous trees in the fragmented Ethiopian highland landscape, particularly confined to church forests. These forests are preserved over centuries by the Ethiopian Orthodox Tewahido Church and are considered sanctuaries for different plant and animal species. However, recent inventories showed that the population of tree species in these forests is not only small but also decreasing over time (Alemayehu Wassie, 2002; Bingelli et al., 2003; Alemayehu Wassie, et al., 2005a). Here we address the following questions: (1) How intense is post dispersal seed predation in church forest, and does this seed predation vary with species and/or habitat, and (2) For how long can tree seeds maintain their viability while buried in forest soils? We hypothesized that there is intense seed predation in church forests, that predation and viability varies with species and that more seeds are predated in forest interior compared to edge and under closed canopy compared to gaps.

2. Materials and Methods

2.1. Study Site and Species

The study was conducted in South Gondar Administrative Zone (SGAZ), Amhara National Regional State, in northern Ethiopia. Based on a large inventory of church forests in the zone a church forest, namely Gelawdios, was selected for the seed predation and seed viability studies reported here (Alemayehu Wassie, 2002; Alemayehu Wassie et al., 2005a). Gelawdios forest is located in 'Dera Woreda' (District) and its first church building was constructed around 1500 A.D. It is located at 11° 38′ N and 37° 48′ E at an altitude of 2549 m. The current forest area covers 100 ha. The mean annual rainfall is about 1216.3 mm (range 1103 -1336 mm), and the average daily temperature is 17 °C. The main rain fall period is between June and September and a lesser amount of rainfall occurs in March and April. Cambisols and Andosols are the predominant soil types in the area (SGDA, unpublished). Based on the survey carried out, the regeneration of trees in this forest was very low and they exhibited no or very few seedlings and saplings in the forest (Alemayehu Wassie, 2002; Alemayehu, et al, 2005a).

To assess the fate of seeds on the forest floor, we focused on post-dispersal seed predation and viability of seeds buried in forest soils. Six and five tree species were selected to assess post-dispersal seed predation and viability of seed, respectively (Table 1). The tree species selected are abundant, have socioeconomic value but regenerate poorly. All of these species had no to very few individuals in the smallest size classes (seedlings and saplings) (Alemayehu Wassie, 2002). Hereafter, the species will be referred to by their generic names.

Table 1. Growth habit, ecology and uses of the study species

Species	Growth Habit	Seed [a] Size (mm)	Habitat/Ecology [a]	Uses [a]
Albizia schimperiana Oliv. (Fabaceae/Mimosoideae)	Tree, Up to 25 m	6 X 5	Dominant in the upper canopy of wet low land or lower highland forests, from 1400-2500 m	Firewood, charcoal, timber, medicine, bee forage, soil conservation and shade
Chionanthus mildbraedii (Gilg & Schellenb.) Stearn (Oleaceae)	Tree, Up to 12 m	13×10	Submontane, montane and upper montane forests, from 1650-2800 m	Firewood, charcoal, timber and farm implements
Ekebergia capensis Sparrm (Meliaceae)	Tree, Up to 30 m	13 X 7	Dominant and co-dominant in upper and intermediate storey of humid upper highland forests, from 1600–3000 m	Firewood, charcoal, poles, timber, farm tools, medicine, bee forage, ornamental and shade tree
Juniperus procera Endl. (Cupressaceae)	Tree, Up to 45 m	5 X 3	A single dominant afromontane forest or more often mixed with *Olea europaea* and *Podocarpus falcatus*.. Commercially known as African pencil cedar, the largest Juniper in the world and one of the two important coniferous trees of Ethiopia, from 1500–3000 m	The wood is termite tolerant and mostly used for fence, posts, poles, timber, particularly used for church building construction, Firewood, medicine, ornamental and shade tree
Olea europaea L [b] (Oleaceae)	Tree, Up to 15 m	7 X 5	Afromontane forest particularly drier highland forests where *Juniperus* is common, secondary scrub, riverine forest, as relic tree on farmland, hardy and resistant once established even in poor soils, from 1250-3000 m	Firewood, charcoal, timber, medicine, tooth brushes (twigs), milk flavoring (smoking wood) walking sticks, farm implements, edible fruit and edible oil
Prunus africana (Hook. f.) Kalkm (Rosaceae)	Tree, Up to 40 m	10 X 7	Widely distributed in humid and semi-humid lower highland forests, humid and semi-humid upper highland forests and humid mountain woodlands, from 1500–2600 m	Firewood, charcoal, poles, timber, medicine (leaves, bark), bee forage, mulch, shade and windbreak

Sources of information: a- Azene Bekele et al. (1993), Demel (1993) and Leggese 1995, Demel Teketay and Granström, 1997; b-Aerts et al. (2006c)

2.2. Experimental design

Post-dispersal seed predation experiment

The post-dispersal seed predation experiment was conducted during the fruiting season of each plant species (March-June, 2004), when considerable seed production and dispersal had occurred (Alemayehu Wassie unpublished data; Sánchez-Cordero and Martínez-Gallardo, 1998). Four habitats, namely closed canopy, gaps, forest interior and forest edge, were identified. Each habitat had four replications and the total number of plots was 16. The plots at the edge were immediately inside the forest edge while the corresponding interior plots were at perpendicular transects to the edge and at distances of 96 ± 8 m (Laurance and Bierregaard, 1997). Sites having above 90% and less than 10% canopy cover were taken as closed canopy and gaps, respectively. Canopy cover was estimated using a Spherical Crown Densiometer. Mature seeds of the study species were collected from several individuals in the forest, mixed well before the experiment and kept similar to naturally dispersed seeds with no fleshy mesocarp.

In each plot, 15 seeds of the study species were randomly selected and arranged in small square subplots (30 X 30 cm) on the ground cleared of litter (Sanchez-Cordero and Martinez-Gallardo, 1998). The seeds were monitored once in a week for 14 weeks. Missed, partially eaten, or pierced seeds were considered as predated. At the end of the experiment all seeds from all plots were counted. The above experimental protocols were used for the six species simultaneously but on separate sub plots to make sure predation of seeds of the study species were independent of one another (16 plots x 6 species x 15 seeds=1440 seeds).

Seed viability experiment

To investigate the long-term behavior of seeds in the soils of church forests, seeds of the five selected tree species were buried in the top-soil of the forest at four locations (replications) following the methods by Demel Teketay and Granström (1997). Mature seeds of the study species were collected from several individuals in the church forests and mixed well before the experiment. All the seeds selected were intact and also checked whether they are not attacked by insects or other organisms. It was assumed that at the time of collection, seeds would have 100% viability. Four different plots (2 x 3.0 m), which were laid out 500 m apart from each other, were selected inside the forests under medium canopy cover of the vegetation to represent average conditions in the forest. Each plot was then divided into three blocks (2 x 0.50 m) with a spacing of 10 cm between them. The blocks were prepared for seed burial by lifting the litter and uppermost five cm layer of soil, with the appropriate care to keep the uplifted part of the forest floor intact and attached to one side of the block. Fifty fresh seeds of each species, except *Olea* (100 seeds) were mixed with soil from the site, to mimic the natural conditions of dispersed seeds, and enclosed in 12 x 20 cm nylon mesh bags (mesh size 1 mm). In each block, one bag of each species was placed in random order on the soil with 30 cm distance between the bags. The lifted litter and soil layer was then put back over the seed bags.

The seeds were buried in March 2005. The buried seeds were exhumed in six months interval i.e. six, 12, and 18 months to assess their viability. In each of these intervals, one complete block was exhumed from each of the four plots (locations) to recover the bags containing seeds of the four species. After exhumation, the bags were opened to examine their contents and the seeds recovered were classified as intact, germinated or non-viable. The seeds that could be easily damaged by the pressure applied on them between two fingers were considered as non-viable. Seeds were considered germinated when living or dead seedlings were encountered in the bags. Seeds with intact internal contents were considered viable after checked by a cutting test (Demel Teketay and Granström, 1997).

2.3. Data Analyses

The data (percentage of seeds predated) was first arcsine transformed to fulfill requirements of parametric test. Then for the Post-dispersal seed predation experiment, two way ANOVA was used to determine effect of species, habitats (Closed canopy vs gaps and forest interior vs forest edge) and interaction effect between species and habitat. Tuckey's post hoc test was used to analyse differences among species (Zar, 1974). Moreover the time effect on predation was analysed by Kaplan-Meier survival analysis and log rank test was used to analyse differences among species and between habitats (Kleinbaum and Klein, 2005).

One-way Analysis of variance was employed to test for significant differences in viability of seeds among species by taking the means of intact and germinated seeds over the study period (18 months). Tuckey's post hoc test was used to compare further differences among species (Zar, 1984).

3. Results

Post-dispersal seed predation

The majority of seeds of the six study species (92 %) which were arranged on the forest floor were predated within 3.5 months. Predation was mainly dependant on the species while habitat had a weaker effect.

Species strongly affected the mean predation percentage ($F_{5, 95} = 11.5$, $P < 0.0001$). Seeds of *Prunus* were completely eaten while those of *Chionanthus* were the least eaten (77.9 %) and the other species were in between (Table 2). Canopy openness affected the predation percentage regardless of species ($F_{1, 48} = 4.5$, $P < 0.05$). Predation was higher in closed canopy plots (93.9 %) than in gaps (87.8 %). Plots in the forest interior (93.8 %) and edges (92.2 %) were not different ($F_{1, 48} = 0.001$, $P = 0.97$). There was no significant interaction effect between species and canopy openness ($F_{5, 48} = 0.46$, $P = 0.8$) and between species and interior vs edge ($F_{5, 48} = 1.7$, $P = 0.14$).

Table 2. Predation of seeds (% ± SE) of the six study species in four habitats after three and a half months. Different letters in same column and row show significant differences at P< 0.05 (Tuckey's post hoc test).

Species	Habitat				Over all species mean
	Closed canopy	**Gap**	**Forest Edge**	**Forest Interior**	
Albizia	95.0 ± 5.2	88.3 ± 5.1	95.0 ± 4.7	90.0 ± 5.4	**92.0 bc**
Chionanthus	76.7 ± 5.3	76.6 ± 5.0	81.7 ± 5.1	76.8 ± 4.9	**77.9 a**
Ekebergia	100 ± 0	93.3 ± 0.7	100 ± 0	100 ± 0	**98.3 bc**
Juniperus	96.7 ± 4.8	86.7 ± 5.2	100 ± 0	96.7 ± 4.8	**95.0 bc**
Olea	95.0 ± 5	81.7 ± 5.2	80.0 ± 5.3	100 ± 0	**89.2 b**
Prunus	100 ± 0	100 ± 0	100 ± 0	100 ± 0	**100.0 c**
Overall habitat mean*	**93.9 b**	**87.8 a**	**92.8 x**	**93.8 x**	

* comparison were made between closed canopy and gaps, and forest edge to forest interior.

Over time the differences in seed survival between species increased (Fig. 1-A). For instance, at the end of the first week, 43% of seeds of *Prunus* were predated while only 6% of the seeds of *Chionanthus* were predated. After four weeks, 87% and

30% of seeds of *Prunus* and *Chionanthus* were predated. The other species were in between (Fig. 1-A). Seed predation rate also varied significantly ($P < 0.05$) between plots in gaps and those under a closed canopy (Fig. 1-B). In the first week, 46% of seeds of all species were predated in closed canopy plots while only 12% of the seeds were eaten in gap plots. After four weeks, 75% and 45% of seeds were predated in closed canopy and gap plots, respectively (Fig. 1-B). Predation rate was not significantly different between forest interior and edge.

In general, based on the amount of seeds predated, the study species could be categorized into two groups: Group I- species with higher seed predation: *Prunus, Ekebergia, Juniperus, Albizia,* and *Olea* and Group II- species with lower seed predation: *Chionanthus* (Table 2). The species could also be ranked according to their rate of seed predation as (in descending order): *Prunus* > *Ekebergia* > *Juniperus, Albizia* , *Olea* > *Chionanthus* (Fig 1-A).

Seed Viability

There were significant differences in the viability of seeds among the study species ($F_{4, 60} = 710.6$, $P< 0.001$). For all species seed viability decreased significantly with burial time except *Juniperus,* which still had 91% of viability after 18 months. Tuckey's *post hoc* test further showed that at the end of the experiment seeds of *Juniperus* had the highest viability, followed by those of *Olea* while seeds of *Albizia, Ekebergia* and *Prunus* had the lowest viability (Table 3). During exhumation, some of the seeds were found already germinated in the nylon bags. At the end of experiment, the percentage of germinated seeds was significantly different among species ($F_{4, 60} = 16.6$, $P< 0.001$). *Albizia* seeds had a high percentage of germinated seeds whereas only few seeds had germinated in the other four species (Table 3).

Although *Albizia* seeds could not maintain their viability in the soil (Fig. 2-A), almost half of the seeds were found germinated with protruding radicles. The loss of viability for this species was considerably due to germination. At the end of the first six months, over 90% of the seeds were found already germinated. However, with increased storage time, the percentage of germinated seeds declined (Fig. 2-A). Seeds of *Ekebergia* and *Prunus* neither germinated nor maintained their viability through out the study period although few germinated seeds of *Ekebergia* were recovered in the first exhumation (Fig. 2-A & B). The seeds had already lost their viability at the end of the first six months (Fig. 2-A2 & A3). The dead seeds of *Prunus* were not recovered after six months since they were completely deteriorated. *Juniperus* seeds maintained their viability substantially through out the study period. It was interesting to note that there were more viable seeds in the second exhumation period than the first and third, though the difference was not statistically significant (Fig. 2-A5). On the other hand, few germinated seeds were recovered in the first and third exhumation (Fig. 2-B5). In the first exhumation date 21% of seeds of *Olea* were found germinated inside the nylon bags. In addition, although non-germinated and non-viable seeds were recovered during the last two exhumation periods, 20% of the seeds of *Olea* were found maintaining their viability at the end of the experiment (Fig. 2-A4 &B4). All germinated seeds of the study species were recovered from the first and third exhumation date except those of *Albizia,* which were found germinated also from the second exhumation (Fig 2).

Table 3. Mean (± SE) proportion (%) of intact, germinated and dead seeds of each study species after 18 months of burial in forest soils. Different letters in the same column show significant differences among species at , $P < 0.05$ (Tuckey's post hoc test).

Species	**Intact Seeds**	**Germinated Seeds**	**Dead Seeds**
Albizia	0.5 ± 0.3 a	47.7 ± 9.7 b	51.8 ± 9.5 b
Ekebergia	0 a	0.5 ± 0.5 a	99.5 ± 0.5 d
Juniperus	91.0 ± 1.2 c	4.7 ± 0.6 a	5.3 ± 0.6 a
Olea	20.0 ± 1.1 b	7.0 ± 3.0 a	73.3 ± 2.9 c
Prunus	0 a	0 a	100 ± 0 d

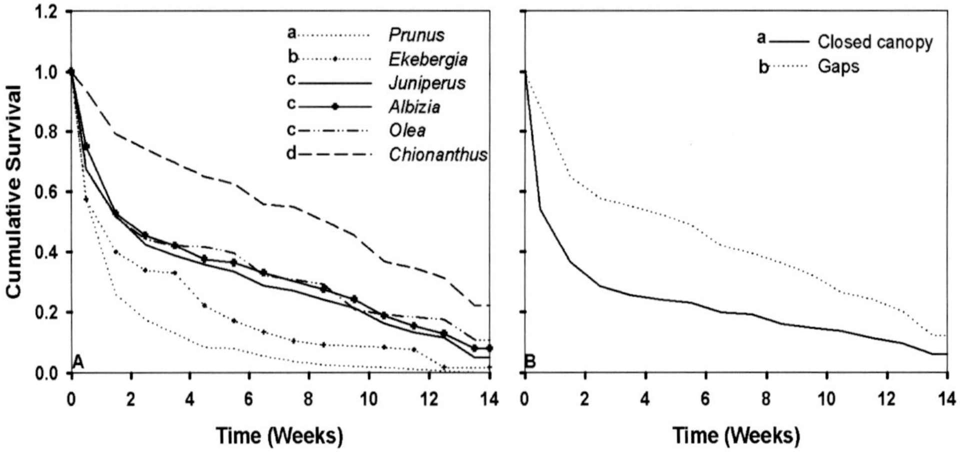

Figure 1. Survival of seeds of six forest tree species showing (A) difference among species and (B) difference between gaps and closed canopy. Different letters show significant differences in the survival of seeds at P< 0.05 (Kaplan-Meier, Log-rank test).

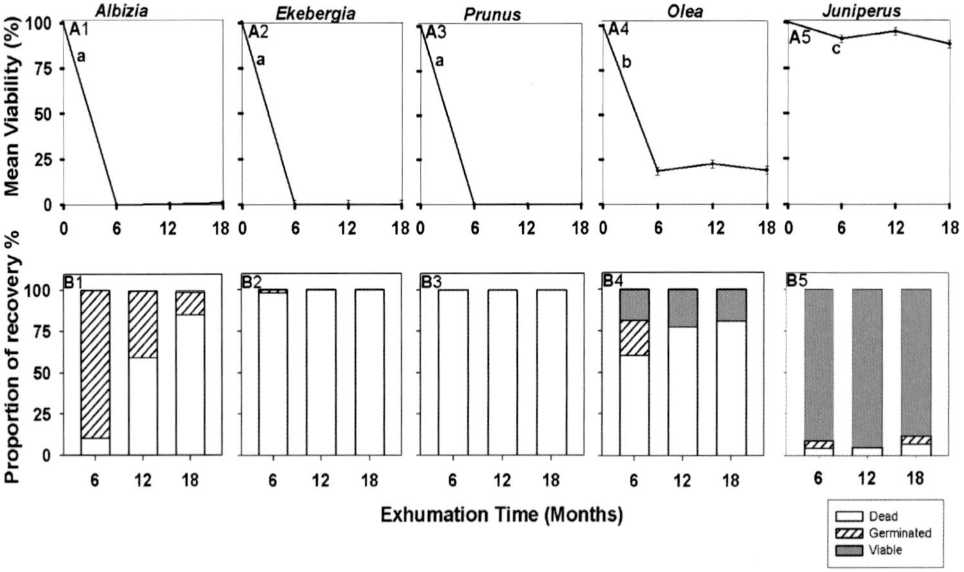

Figure 2. (A) Mean Percentage of seeds recovered intact (B) Proportion (%) of intact, germinated and dead seeds of the five species across the exhumation dates. Small letters shows differences between species (Tuckey's post hoc test). Different letters shows significant differences at P<.05

4. Discussion

Post-dispersal seed predation

Once dispersed, seeds may be destroyed by a variety of animals and microorganisms (Hulme and Borelli, 1999). When regeneration is limited by seed availability the consequence of seed predation for species regeneration and population growth may be very significant (Crawley, 1992; Fenner and Kitajima, 1999). Most of the seeds were found eaten on the spot as observed in our study from left over pieces, gnawed and pierced recovered seeds. Seeds removed from the spot by animals were probably either placed in deep burrows (Smith & Reichman 1984; Montgomery & Gurnell, 1985), consumed later somewhere or scatterhoarded. Successful establishment is not expected from deep burrows, and survivorship of seeds from scatterhoards is generally low (Jensen 1982, 1985; Jensen & Nielsen 1986). Studies on different species showed that only 0.02-10% of the removed seeds were able to establish seedlings (Jansen et al., 2002; Hulme, 2002). We, therefore, considered that removal was almost part of predation (Sullivan, 1979; Hulme & Borelli, 1999).

The substantially higher percentage of seed predation (92 %) in our result indicates that post-dispersal seed predation in church forest could significantly reduce the availability of tree seeds for regeneration. Hulme (2002) reviewed several field studies and showed that post-dispersal seed predation is often much lower than 90%. This review also showed a significant positive correlation between rates of pre- and post-dispersal predation (r=0.857). For our forest this may imply that the high post-dispersal predation might have been preceded by higher pre-dispersal predation. Predation, both amount and rate, was mainly dependant on species. One of the crucial factors determining seed fate is the chemical composition of the edible portion (Passos and Oliveira, 2003). As previous studies indicated, most animals showed marked feeding preferences for fruits and seeds with a high energy and/or nutritional content, and avoided those containing toxic compounds (Janzen, 1971, 1986; Smythe 1986; Martínez-Gallardo & Sánchez-Cordero 1993; Sánchez-Cordero and Martínez-Gallardo, 1998). Accordingly, seeds with higher nutritive values tend to attract more predators while those species that have toxic contents attracts less predators.

Predation was also dependant on canopy openness since more predation was encountered in closed canopy plots than those in gaps, as expected. Sánchez-Cordero and Martínez-Gallardo, (1998) also found similar results that post-dispersal fruit and seed removal was habitat-dependent for most plant species tested, with lower removal in gaps compared with that under storey of mature forests. The possible explanation for higher predation under closed canopy could be that most predators prefer a specific habitat to obtain sufficient nutrients for survival and reproduction while avoiding being eaten by other foraging animals (predators) (Kelt et al., 2004). Especially rodents and other small mammals avoid the risk of being exposed to their predators in gaps while they eat the seeds. On the other hand, in our study forest interior vs edge did not show a significant effect as expected, neither in amount nor in rate of predation. Kollmann and Buschor (2002) and Fleury and Galetti (2004) reviewed previous studies in other areas and reported different results with various justifications. For instance, Laurance (1994) and Malcolm (1994) suggested that seed predation at the forest edge should be higher than in the forest interior because of the high diversity of small mammals in this microhabitat. However, Burkey (1993) observed that seed predation was reduced closer to the forest edge due to the lower density of rodents in this microhabitat, caused by an increase in the density of meso-predators. Also non-significant differences in seed predation between forest interior and edge were reported (Notman et al. 1996; Holl and Lulow, 1997; Wong et al. 1998), which supports our findings.

Seeds of *Prunus* and *Ekebergia* were not only entirely predated but very rapidly. During field work, birds, rodents and mammals (monkeys and apes) have been observed heavily consuming seeds of *Prunus* and *Ekebergia* (Alemayehu Wassie, per.obs.). Since seeds of *Prunus* species do not tolerate desiccation (Legesse Negash, 2004; Sacande, et al. , 2004), animals consume them before they either germinate or rot. The fact that *Albizia* and *Juniperus* exhibited relatively less predated seeds may be attributed to their hard seed coats. Similarly, seeds of *Chionanthus* and *Olea*, belonging to same Oleaceae family had relatively much lower seed predation, possibly due to the low nutritive value of their seeds, the hard seed coats, or both. Aerts et al. (2006b) reported much lower percentages (3.2-7%) of eaten seeds of *Olea* within one month in a drier shrubland of northern Ethiopia, suggesting that seed predation is severe in church forests of northern Ethiopia. This maybe the result of higher predator densities, as these forests are refuges for many birds, mammals, rodents and insects (Alemayehu Wassie, et al., 2005a). During fieldwork, many granivorous birds *(Speckled Pigeon; Dusky Turtle Dove; Helmeted Guineafowl; Yellow-necked Francolin/Spurfowl; Village Weaver; Yellow-crowned Bishop)* and mammals and rodents (*miniature deer; Menilek's bushbuck; monkeys; apes; porcupines; rabbits; squirrels* and *mice)* were observed in the study and other church forests (Alemayehu Wassie, per.obs.) that could potentially contributed to the observed seed predation.

Hulme and Borelli (1999) hypothesized that seed predation has considerable effect on regeneration if: (1) trees do not regenerate by vegetative means primarily; (2) there is no large persistent seed bank that buffers seed losses to predators; (3) seed predators are not satiated by mast seed crops and/or (4) regeneration is seed-limited rather than microsite-limited. For instance, seed predation by the deer mouse and other rodents has contributed to the failure in regenerating cutover forest lands at the University of British Columbia Research Forest (Sullivan, 1979). Others indicated that even if there is intense post-dispersal seed predation, its effects may be minor in the regeneration of trees (Louda 1989; Hulme 1996a&b).

In our forests seed availability was one of the limiting factors for regeneration of *Ekebergia, Juniperus, Olea* and *Prunus* (chapter 6), and these species (including *Albizia*) did not accumulate large persistent soil seed banks in Gelawdios (Chapter 3, Alemayehu Wassie and Demel Teketay, 2006). By combining these two evidences with earlier information (Azene Bekele et al., 1993) that these species do not mainly depend on vegetative reproduction, post-dispersal seed predation may have considerable influence on seed availability for successful establishment of these species in church forests (Crawley, 1992; Fenner and Kitajima, 1999). Moreover, the higher predation pressure under the closed canopy may aggravate the problem, particularly, of *Prunus* that prefers to germinate in shady plots in the forest (Chapter 6). Seed predation not only undermines the availability of seeds for the regeneration of these species but it may also lead to a negative impact on the genetic make up and vigour of the population since predators selectively consume the high quality seeds (Benkman, 1995; Fenner and Kitajima, 1999). In the case of *Chionanthus,* large number of seeds were found in soil seed bank analysis although not all of them viable (Chapter 3, Alemayehu Wassie and Demel Teketay, 2006), suggesting that the current low predation of seeds soil seed bank may not have a critical effect on seed availability for its regeneration as compared to the other five species.

Seed viability

Maintaining seed viability in soils for a longer time determines the ability of the seeds to form a persistent soil seed bank. The main physiological factor that governs seed viability in the soil is the presence of dormancy that inhibits immediate germination even under favorable conditions (Fenner and Kitajima,

1999). The ability to resist microbial infestation and degradation is another factor. These factors are in turn, determined by the genetical and phenotypical nature of the species (Vazquez-Yanes and Orozco-Segovia, 1993). Based on the viability of seeds after burial of 18 months in the soil, the study species can be categorized into four groups: (1) species that maintain their viability significantly throughout the period, i.e. *Juniperus*; (2) species that only weakly maintain their seed viability, i.e. *Olea;* (3) species that do not maintain their seed viability (lost almost half of their viability due to germination), i.e. *Albizia*: and (4) species that can not maintain their seed viability due to natural death immediately following burial in the soil, i.e. *Ekebergia* and *Prunus*.

Seeds of *Juniperus* possess dormancy (Laurent and Chamshama, 1987; Demel Teketay and Granstrm, 1996; Eshetu Yirdaw and Leinonen, 2002) and have been able to maintain their viability. In the first and third exhumation dates (August 2005 & 2006), which corresponded to the raining season of the area, germinated seeds were recovered in the nylon bags. This may indicate that seeds could loose their viability overtime whenever moisture is available. Demel Teketay and Granström, (1997) showed also *Juniperus* viability decline from 100 to 37% over 4 years. Although *Olea* possesses some level of dormancy (Demel Teketay and Granström, 1997; Aerts et al., 2006c), the low percentage of viable seeds recovered from the soil suggests that seeds of this species have little possibility of maintaining viability in forest soils for longer time. Demel Teketay and Granström, (1997) reported also that even if there were intact seeds recovered after 4 years these species would be expected to have transient soil seed banks. Even though *Juniperus* and *Olea* can form transient soil seed banks, their seeds were not found in soil seed bank assessment in same forest (Chapter 3, Alemayehu and Demel, 2006). This could either be due to the absence of a continuous supply from seed rain or to the severe seed predation.

Although seeds of *Albizia* possess seed coat imposed dormancy (Demel Teketay, 1996), they can not maintain a soil seed bank because they tend to germinate whenever there is enough moisture. Seeds of *Ekebergia* and *Prunus* did not only lack dormancy but also did not withstand microbial degradation in the forest soil. Many *Ekebergia* and almost all *Prunus* seeds were recovered completely deteriorated and rotten. Removal of litter improved the germination of these two species (Chapter 6), maybe associated with a reduction of microbial and fungi attack by litter removal (Kitajima and Fenner, 2000; Abdella Gure, 2004). These two species require safe sites (Sensu Harper, 1977 e.g., bare mineral soil, exclsouers from animals and favorable microclimate conditions) and germinate immediately on the forest floor. Similarly Demel Teketay and Granström, (1997) reported that *Ekebergia* did not accumulate a soil seed bank. All investigated species in this study were not found to have persistent soil seed bank in the previous soil seed bank assessment done on same site (Chapter 3, Alemayehu and Demel, 2006).

Implications for regeneration

The predation experiment period (March-June) was the high seeding season for most trees (Alemayehu Wassie, unpublished data). End of June is the beginning of the main rainy season of the area. Most of the seeds produced need to wait at least the first shower to germinate. But according to our result seeds face severe predation before the rain starts. Results from the seed predation and viability experiments indicate that the fate of seeds of *Ekebergia* and *Prunus* is complicated. They suffered from higher and rapid predation and also whenever there is a chance to be buried there is high risk of microbial deterioration. This might, ultimately, limit the number of available seeds for regeneration. Seeds of *Albizia,* also can not wait to be buried in the soil but on surface they face predation. Seeds of *Juniperus* and *Olea* also suffer from high predation on the forest floor, but they can maintain viability of some of the seeds until they get favorable conditions. The annual seed input from the seed rain for these species in the same site was for *Albizia* = 4.2, *Olea* = 4.7, *Ekebergia* = 5.4, *Juniperus* = 11.8 and *Prunus* = 22.9 seeds/m^2/yr (Alemayehu Wassie, unpublished data), which is low, taking in to account that in our experiment almost all seeds are predated. In general, our results provide empirical evidence that unavailability of seeds associated with seed predation and short longevity of seeds in combination with a low seed input from the seed rain may limit regeneration of tree species in church forests. On the other hand, for *Chionanthus* with higher seed numbers in seed rain (116.3 seeds/m^2/yr, Alemayehu Wassie, unpublished data), having a soil seed bank (Chapter 3, Alemayehu Wassie and Demel Teketay, 2006), taking into account 22% of the seeds are not removed in our experiment, seed predation may not have a significant impact on seed availability. In order to maintain those species with higher predation rate and short viability in soil, the standing vegetation of church forests needs to be sustainably managed and conserved. Not only the presence of sufficient adult trees are needed but also the seed producing capacity (quantity and quality) of these populations need to be taken into account. Additionally sowing of seeds and planting of seedlings can help, especially when combined with active reduction of predation and herbivory (e.g. in the form of enclosures (Fleury, and Galetti, 2006) and clumping alternative food sources (Sullivan, 1978 & 1979)). Similar studies on more species from different church forests are required to determine their future fate and design appropriate mechanisms to promote their sustainable management and conservation.

CHAPTER 5

THE EFFECT OF LIVESTOCK ON TREE REGENERATION IN CHURCH FORESTS OF ETHIOPIA

Alemayehu Wassie, Frank Sterck, Demel Teketay and Frans Bongers

Abstract

In northern Ethiopia forests around churches are the last remnant forest patches. These forests are currently under threat, probably due to diminishing areas and extensive grazing by cattle. We assessed the effect of livestock grazing on the regeneration of four indigenous tree species that have high abundance and socioeconomic value but limited regeneration in two forests and focused on seed germination, seedling survival and growth in relation to livestock trampling and grazing.

Livestock grazing has a clear and strong negative effect on germination, seedling growth and mortality. In fenced plots, more seeds germinated, seedling survival was higher and seedlings grew faster. Seed germination was higher inside the forest than in the adjacent open area for all species. Seedling survival was not different between forest interior and open field, except for unfenced plots in the open field where survival was lower because of the higher grazing pressure. In unfenced plots, no seedlings survived until the end of the year (both in forest interior and open field), indicating that the grazing effect ultimately did not differ. Mortality rates, however, were higher in the open field. The significant interaction between fencing and species on seed germination and seedling survival revealed that the magnitude of damage due to grazing can vary with species.

We conclude that for effective indigenous tree species regeneration in these church forests the control of livestock pressure is mandatory. Seeds dispersed outside the forest will not have a chance to establish seedlings, grow and colonize the surroundings. Livestock grazing thus has a paramount impact on the long-term sustainability of church forests and their role in restoring the degraded surroundings.

1. Introduction

Many of dry tropical forests have been fragmented into small patches, and forest structure and regeneration have been influenced due to this fragmentation and habitat loss (Cabin et al., 2002). Human induced disturbances mainly via e.g. grazing activities or tree harvesting strongly influence the regeneration success of woody species and in turn determine the vegetation structure and composition of these forests (Cotler and Ortega-Larrocea, 2006). As a result the persistence of the remnant forest patches and their indigenous species in many areas are threatened. Here, we report on the regeneration of indigenous tree species in the fragmented Afromontane landscape in Ethiopia. In northern Ethiopian highlands, the forests around churches (hereafter referred to as church forests) are almost the last remnants of the original forest cover, and thus these forests are of paramount importance for forest development and restoration. However, previous pilot studies in church forests showed that most of the tree species had small population sizes and some species appear to have no regeneration (Alemayehu Wassie, 2002; Bingelli et al., 2003; Alemayehu Wassie et al., 2005).

The role of livestock grazing might be either facilitative or hamper regeneration. It has been suggested that a light level of grazing can increase tree regeneration by removing competitive vegetation, remove fire hazard and fertilizing by droppings. Large herbivores can also create patches of bare ground through grazing and trampling. These disturbances may create "safe sites" (sensu Harper, 1977) for seeds to germinate in 'regeneration niches' (González-Hernández and Silva-Pando 1996; Mayle 1999; Vallentine 2001; McEvoy et al., 2006). On the other hand, continuous and intensive grazing may cause irreversible damage (Opperman and Merenlender, 2000). Livestock can compact soil, exacerbate erosion, consume and trample tree seeds, seedlings and browse saplings, and thereby preventing forest regeneration. Eventually, the tree canopy thins out and disappears (Scowcroft, 1983; Stone, 1985; Blackmore and Vitousek, 2000; Fleischner, 1994; Opperman and Merenlender, 2000; Ramirez-Marcial et al., 2001; Carolina and Javier, 2001). Grazing by animals can reduce forest productivity through a reduction in seedling growth rate and survival, and by changing tree form (Wilkinson and Nielsen, 1995; Bulinski and McArthur, 1999; Miller et al., 2006).

Three main components of structural change have been identified in a grazed forest: browsing on seedlings which limits stem density; browsing on leading shoots which limits height growth; and browsing on side shoots and climbers which reduces foliage density (Gill and Beardall 2001; McEvoy et al., 2006). There is, however, great variability in the effects of grazing on plant population dynamics, depending on factors such as the susceptibility of the woody species involved (Sun et al., 1997; Garin et al., 2000; Motta, 2003; Teich et al., 2005). According to Miller et al. (2006) protection given to the relatively palatable seedlings by neighboring tall, unpalatable thistles is consistent with two ecological hypotheses: the associational plant refuge hypothesis (Pfister and Hay, 1988) and apparency theory (Feeny, 1976). Associational refuges are most commonly described in terms of palatability of a particular plant relative to that of the neighboring vegetation. Associational refuges may also operate through other influences on herbivore foraging behavior such as searching effort (Milchunas and Noy-Meir, 2002). In this sense, the capacity to hide seedlings may be as pertinent to offering "refuge" as being unpalatable. This effect can be linked directly to the concept of plant appearance (Feeny, 1976). An unapparent plant is one that is hard to find, which in turn can be influenced not only by factors such as its size, growth form, secondary chemistry and persistence, but also by its relative abundance within a community, characteristics of neighboring plants, and herbivore characteristics (Feeny, 1976; Miller, et al., 2006; Smit et al., 2006).

In most church forests domestic animals graze and browse, and are sheltered under the trees to escape from strong sunrays and storms when the outside landscape was bare (Alemayehu Wassie, 2002; Alemayehu Wassie et. al, 2005). Therefore, livestock-induced disturbances are expected to be among the major factors constraining regeneration of woody species and contributing, ultimately, to the decline of woody species populations in the Ethiopian church forests.

Another important factor is light which has a paramount effect on the regeneration of tree species. In general, light may limit regeneration when moisture is not limiting (e.g. in the wet season), suggesting that regeneration may be facilitated in the more open sites. On the other hand, when soil moisture is mainly limiting (e.g. in dry season), regeneration may be facilitated more in shady sites like in forests (Denslow, 1980; Fetcher et al., 1985; Vitousek and Denslow, 1986; Belsky et al., 1993; Gerhardt, 1996; Chen et al., 1999; Bullock, 2000; McLaren and McDonald, 2003; Asbjornsen et al., 2004).

The present study aims to assess the effect of livestock grazing on seed germination, seedling survival and growth of indigenous tree species in Ethiopian church forests. We set out to test whether livestock exclusion would increase regeneration of indigenous trees in church forests. We addressed the following specific questions: (1) How does grazing affect regeneration of indigenous tree species? and (2) Is the effect dependent on whether the seedlings are inside the forest or in the open field?

We hypothesized that livestock grazing reduces regeneration of indigenous tree species (Jimenez et al., 2005; Teich et al., 2005; McEvoy et al., 2006), that the magnitude of the reduction depends on the tree species involved (Sun et al., 1997; Garin et al., 2000; Motta, 2003, Teich, et al., 2005), and that being inside the forest dampens the effects (Alvarez-Aquino et al., 2004).

2. Material and Methods

2.1. Study Site

The study was conducted in South Gondar Administrative Zone (SGAZ), Amhara National Regional State, in northern Ethiopia. Based on a large inventory of church forests in SGAZ (Alemayehu Wassie et al., 2005; unpublished data) two forests, Dengolt and Gelawdios, were selected for the study. According the survey carried out, the status of regeneration of the different trees in these forest was very low (Alemayehu Wassie, 2002;). Dengolt is a forest around a church built in 1300 AD. It is located in 'Estie Woreda (District)' at 11° 36′ N and 38° 04′ E and at an altitude of 2500 m. The current forest area covers 25 ha. Gelawdios is a forest around a church built in 1500 AD and is located in 'Dera Woreda' at 11° 38′ N and 37° 48′ E at an altitude of 2549 m. The current forest area covers 100 ha.

The two sites share similar agro-climatic conditions and were at 20 km distance form each other. The mean annual rainfall is about 1216 mm, and ranges from 1103 to 1336 mm. The average daily temperature is 17 °C. Cambisols and Andosols are the predominant soil types in the area (SGDA, unpublished). In both forests livestock grazing has been observed although the church forbids the practice.

2.2. Study species

We selected four tree species that have high abundance and socioeconomic value but limited regeneration (Table 1): *Juniperus procera* Endl., *Ekebergia capensis* Sparrm, *Prunus africana* (Hook. f.) Kalkm and *Olea europaea* L. subsp. *cuspidata* (Wall. ex DC.) Cifferri (hereafter referred to by their genus names). We focused on three aspects of the regeneration of the species selected, namely seed germination, seedling survival and growth.

Table 1. Growth habit, ecology and uses of the study species.

Species	Growth Habit	Seed Size[a] (mm) and Viability	Habitat/Ecology [a]	Uses [a]
bergia capensis rrm liaceae)	Tree, upto 30 m	13 X 7, Short viability	Dominant and co-dominant in upper and intermediate storey of humid upper highland forests, from 1600–3000 m	Firewood, charcoal, poles, timber, farm tools, medicine, bee forage, ornamental and shade tree
iperus procera Endl. pressaceae)	Tree, upto 45 m	5 X 3, Longer viability	A single dominant afromontane forest or more often mixed with *Olea europaea* and *Podocarpus gracilior*. Commercially known as African pencil cedar, the largest Juniper in the world and one of the two important coniferous trees of Ethiopia, from 1500–3000 m	The wood is termite tolerant and mostly used for fence, posts, poles, timber, particularly used for church building construction, Firewood, medicine, ornamental and shade tree
a europaea L. subsp. *pidata* [b] eaceae)	Tree, upto 15 m	7 X 5, Intermediate viability	Afromontane forest particularly drier highland forests where *Juniperus* is common, secondary scrub, riverine forest, as relic tree on farmland. Hardy and resistant once established even in poor soils, from 1250-3000 m	Firewood, charcoal, timber, medicine, tooth brushes (twigs), milk flavoring (smoking wood) walking sticks, farm implements, edible fruit and edible oil
nus africana (Hook. f.) km saceae)	Tree, upto 40 m	10 X 7, Short viability	Widely distributed in humid and semi-humid lower highland forests, humid and semi-humid upper highland forests and humid mountain woodlands, from 1500–2600 m	Firewood, charcoal, poles, timber, medicine (leaves, bark), bee forage, mulch, shade and windbreak

Sources of information: a = Azene Bekele et al. (1993), Demel Teketay (1993), Leggese Negash (1995), Demel Teketay and Granström, 1997 ; b = Aerts et al. (2006c)

2.3. Experimental design

In each of the two church forests, six plots (20 x 10 m) were selected randomly at different places of the forest. Each plot was divided into two blocks (10 x 10 m) of which one was fenced. The fence was made by wooden poles and barbed wire, and was designed to exclude only livestock and other bigger mammals in the forest. Each block was subdivided in two sub-blocks (5 x 10 m) of which one was used for the germination experiment and the other for the seedling survival and growth experiments. The experimental design was a split plot design with randomized plots for the fixed factors, i.e. fencing, four species and two forests.

For the seed germination experiment, forty seeds of each species were sown (end of June, 2005) at 10 cm intervals between seeds and rows at a depth equivalent to their seed diameter (Hewitt and Kellman, 2004) in each block (giving a total of 6 plots x 2 blocks x 4 species x 40 seeds = 1920 seeds). The positions of seeds were marked with wooden stick picks to facilitate follow-up of the experiment. Emergence of seedlings above the soil was monitored every month for one year, and seedlings were counted and tagged to avoid double counting.

For the seedling survival and growth experiment, 10 seedlings (<1 year old) for each of the four species were planted in each block (giving a total of 6 x 2 x 4 x 10 = 480 seedlings). The seedlings were raised in a nearby nursery prior to the experiment. The nursery has similar climatic conditions and soil types as the forests and was located 5 km from Dengolt. The seedlings were exposed to full sunlight two months before planting time for hardening. After planting (end of June, 2005) the total initial plant height was recorded (from ground till main apex) and the averages were 9.4 cm ± 4.3 for *Juniperus,* 8.7 cm ± 3.4 for *Ekebergia,* 13.1 cm ± 3.9 for *Prunus*, and 18.5 cm ± 5.9 for *Olea*. The survival and height of seedlings were recorded every month for twelve months, i.e. until next rainy season.

Additionally, the effect of forest cover (inside forest versus open field) on the fencing treatments was assessed only in the Gelawdios forest due to logistics reasons. Six additional plots were selected outside the forest (close to the forest edge) and subdivided into two blocks as was done in the forest interior. The same protocol was followed for seedling germination and for growth and survival.

2.4. Data Analyses

The three regeneration aspects seed germination, seedling survival and seedling growth were analyzed separately.

Seed germination

To analyse the effect of fencing (and thus livestock exclusion) inside the forest on seed germination, the data from the two forests were used. Forest (or site), fencing and species were used as independent factors in a univariate analysis of variance, and number of seeds germinated as respondent variable according to split plot design. Number of seed germinated was log transformed, Log (number+1), to fulfill the assumptions of parametric test (normality and homogeneity of variances). Differences among species were tested using Tukey HSD *post hoc* tests (Zar, 1984).

For analysis of the effect of forest cover (i.e. comparison of effects inside the forest with open field), only the data from Gelawdios forest were used. Here fencing, species and site (inside or outside of forest) were used as factors in a univariate analysis of variance, and number of seed germinated, after being log transformed, was the respondent variable. Again, differences among species were tested using Tukey HSD *post hoc* tests.

Seedling survival and growth experiment

The seedling survival over time was analyzed using a time to event approach, which measures the time to an event for each seedling (Altman and Bland, 1998). In this study, the critical event was death of seedlings. The model used for the analysis was Kaplan-Meier Survival Analyses (Bland and Altman 1998). Equality of survival distributions for factors, i.e. forests, fencing and species was tested using log-rank test (Kleinbaum and Klein, 2005). For the site comparison fencing, species and site were used as factors. Cox-regression of survival analysis was used to see the interactions among factors explicitly (Kleinbaum and Klein, 2005).

Effects of fencing on seedling growth was analysed for a whole year (12 months). Additionally, the first three months were analysed since mortality was high after 12 months, especially in unfenced plots. The change in growth over time was assessed by calculation of relative growth rate (RGR), which expresses growth in terms of a rate of increase in size per unit of initial size ((Evans, 1972; Hunt,1982). The formula used to calculate RGR was: $RGR = (\log_e H_n - \log_e H_o)/ (t_n - t_o)$, where H_n- height measured at specified time; H_o - initial height measurement; $(t_n - t_o)$- the unit time differences between specified time. RGR accounts for differences in initial seedling height and better depict seedling growth potential (Land and Rieske, 2006). The RGRs distribution meets the requirements of parametric test (normality and homogeneity of variances), and in a univariate analysis of variance, forest, fencing and species were incorporated as factors. For the site comparison, fencing, species and site were used as factors. All statistical analyses were done using SPSS Version 12.0.1 (Field, 2005).

3. Results

3.1. Seed germination

Fencing and species significantly affected the mean number of germinated seeds while forest had no effect (Table 2 A). In fenced plots germination was always higher than in unfenced ones (Fig. 1). The number of germinated seeds was lowest for *Juniperus,* highest for *Prunus* and intermediate for *Ekebergia* and *Olea* (Fig. 1). There was a significant interaction effect between forest and species as well as between fencing and species on the germination of seeds, but effects of forest and fencing, and forest, fencing and species did not interact (Table 2 A, Fig. 1).

There was a significant effect of site, fencing and species on the mean number of germinated seeds (Table 2 B). Inside the forest the number of germinated seeds was higher than outside (Fig. 1) and fenced plots had more germinated seeds. *Juniperus* had the lowest germination, *Prunus* the highest and the other two species were intermediate (Fig. 1). Fencing significantly interacted with species, but site and fencing, site and species, and site, fencing and species did not interact (Table 2 B, Fig. 1).

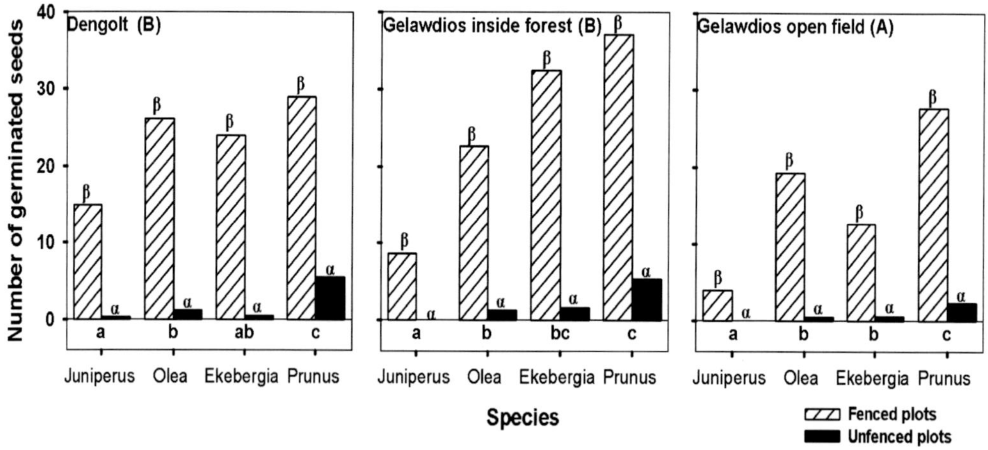

Figure 1. Effects of forest/site, fencing and species on the number of germinated seeds of the four study species. Bars show mean values, capital letters differences between forests/sites, small letters differences among species, greek letters differences between fenced and unfenced plots and different letters significant differences (p<.05)].

3.2. Seedling survival

Forest, fencing and species significantly affected the survival of seedlings (Kaplan-Meier survival analysis, Table 4). Seedlings survived more in Dengolt (37.5%) than Gelawdios (19.4%). Seedling survival was significantly higher in fenced plots (56%) than in unfenced ones (0%) and it was also significant for all species (Table 4, Fig. 2). Seedling survival of the four species was significantly different from each other, both in the fenced and in the unfenced plots (Table 4). Further pairwise comparison showed that *Olea* had the highest seedling survival (93.3%), *Juniperus* the lowest (37.5%) and the other two species were intermediate in the fenced plots. In the unfenced plots, however, *Ekebergia* stayed longer (3.5 months) though no seedlings survived till the end of the year, while the other three species died quickly (< 2.5 months) confirms an interaction effect of species and fencing (Table 3, Fig 2). The effect of forests and fencing also interacted significantly that seedling survival differed significantly in fenced plots of the two forests (Dengolt being higher than Gelawdios) but it remained similar in the unfenced plots (Table 3). Although the difference among species varies between the forests (Fig. 2), the interaction effect of forests and species was not significant in seedling survival (Table 3).

The seedling survival did not significantly differ between forest interior (19.4%) and open field (24%, Table 3). However, in the unfenced plots the survival rate inside the forest was higher than in the open field, as shown by the interaction between fencing and site (Table 5, Fig 2). The results also showed that species and site do interact. Inside the forest, *Olea* had the highest seedling survival, and *Juniperus* the lowest, while *Prunus* and *Ekebergia* were intermediate, although differences were not always significant (Fig. 2). In the open field, on the other hand, seedling survival of *Juniperus* was intermediate. *Prunus* and *Ekebergia* did not differ, but both had significantly lower seedling survival than *Olea* (Fig. 2).

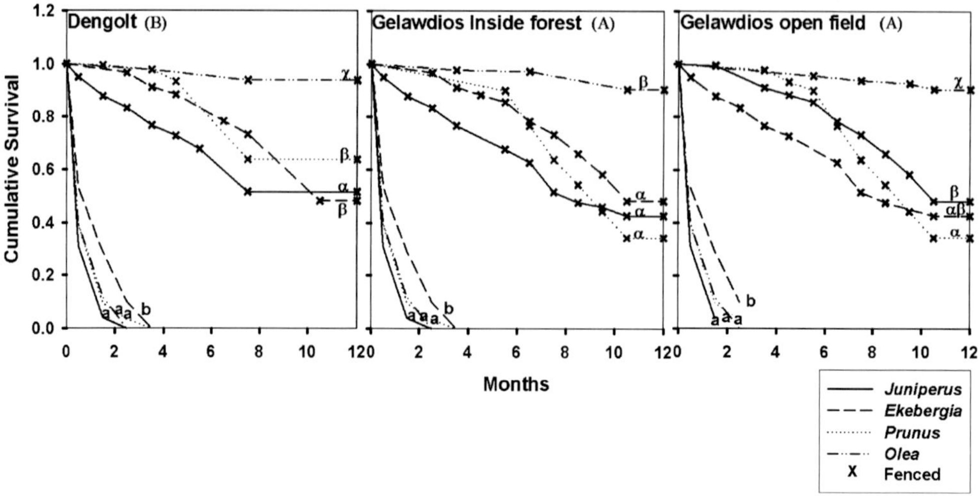

Figure 2 Seedling survival curves for forest/site, fencing and species. Capital letters show differences in seedling survival between forests/sites, small letters differences among species in unfenced plots, greek letters difference among species in fenced plots and different letters significant differences (p<.05). In all cases the difference between fenced and non-fenced plots are significant (Kaplan-Meier survival analysis, p<0.05)].

3.3. Seedling growth

Fencing significantly affected height growth of seedlings: fenced seedlings grew better than unfenced ones, both during the first three and over the twelve months period (Table 2 A). In unfenced plots, only *Ekebergia* and *Prunus* survived the first three months (albeit with retarded height growth) while no seedlings survived after twelve months (Fig. 4). The forest did not affect RGR but species were different both during the first three and twelve months. Forest and species significantly interacted, but there was no interaction between forest and fencing and among forest, fencing and species in the first three months. Interaction effects after 12 months could not be analysed since seedlings did not survive (Fig. 4).

Site did not affect the relative seedling height growth (Table 2 B). In the open field, no seedlings survived in the unfenced plots and, as a result, fencing interacted with site (Fig. 4). Though species were different, no interaction was found with site. *Olea* seedlings showed a retarded growth in fenced plots in the open field, possibly indicating grazing/browsing by animals.

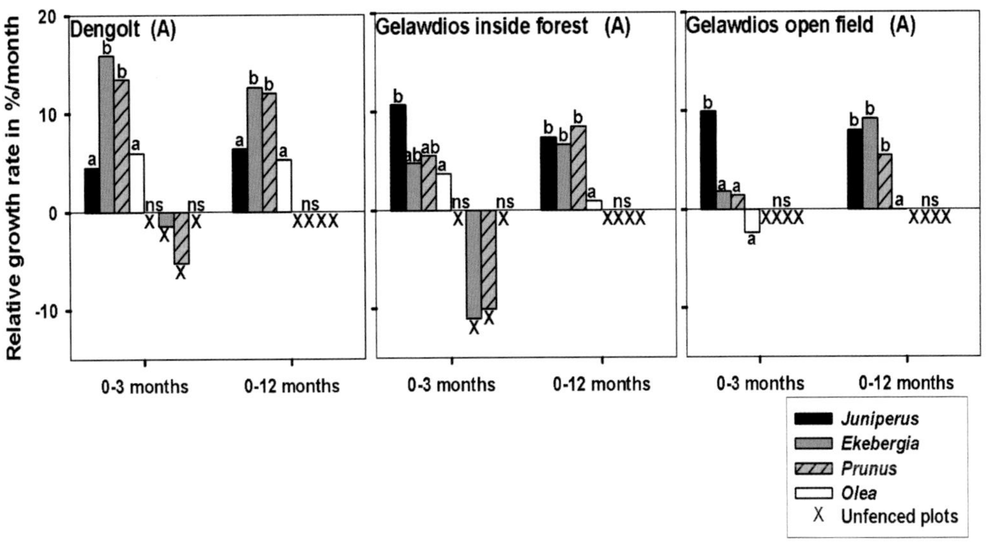

Figure 4 Effects of forest/site, fencing and species on the relative height growth rate (RGR) of seedlings of the four study species. Bars show mean values (RGR/month), capital letters differences between forests/sites, small letters difference between species in fenced plots for each growth period and different letters significance difference ($p<.05$)]. For *Ekebergia* and *Prunus*, there were significant differences between fenced and unfenced plots ($p<0.05$) during first three months only and ns refers to no surviving seedlings for analysis.

Table.2 (A) Effects of forest (between forest), fencing and species with their interactions (B) Effects of site (inside vs outside forest), fencing and species with their interactions on number of germinated seeds and relative growth of seedlings. Mean values of contrasts and ANOVA results with F statistic are shown. Different letters show that means are significantly different, within contrasts (ANOVA, with in the case of species a Tukey posthoc test). Level of significance: *** = P < 0.001; ** = P < 0.01; * = P <0.05; ns = non-significant).

A

Dependent variables	Main Effects											Interactions		
	Forest (F)			Exclosure (E)			Species (Sp)					F*E	F*Sp	E*Sp
	Dengolt	Gelawd.	Statistic	Fenced	Unfenced	Statistic	Juniperus	Olea	Ekebergia	Prunus	Statistic	Statistic	Statistic	Statistic
Germinated seeds	12.8 a	13.7 a	F= 0.01 ns	24.4 b	2.0 a	F= 4561 ***	6.0 a	12.8 b	14.7 b	19.3 c	F=24.2 ***	F=0.04 ns	F=4.3 *	F=5.7 **
RGR (0-3)	7.1 a	3.8 a	F=0.5 ns	8.2 b	-6.2 a	F=23.6 ***	7.4 b	4.8 a	8.5 b	9.2 b	F=3.9 *	F=9.8 ns	F=6.8 ***	F=36.5 ns
RGR (0-12)	8.6 a	5.4 a	F=3.5 ns	7.3			6.7 b	3.1 a	11.1 c	11.4 c	F=12.4 ***		F=2.9 ns	

B

Dependent variables	Main Effects											Interactions		
	Site (S)			Exclosure (E)			Species(Sp)					S*E	S*Sp	E*Sp
	Forest	Open field	Statistic	Fenced	Unfenced	Statistic	Juniperus	Olea	Ekebergia	Prunus	Statistic	Statistic	Statistic	Statistic
Germinated seeds	13.7 b	8.4 a	F= 9 *	20.6 b	1.5 a	F= 1804 ***	3.2 a	10.9 b	11.9 b	18.1 c	F=68.3 ***	F=0.47 ns	F=0.71 ns	F=3.4 *
RGR (0-3)	6.1 a	3.1 a	F=3 ns	4.5			10.4 b	1.2 a	3.4 a	3.5 a	F=6.1 **		F=0.2 ns	
RGR (0-12)	5.4 a	4.9 a	F=0.04 ns	5.2			7.8. b	0.4 a	8.2 b	6.8 b	F=41.6 ***		F=0.7 ns	

Table 3. Effects of forest/site, fencing and species on survival of seedlings of the four study species. Log rank test statistic for main effects (Kaplan-Meier survival analysis) and Wald statistic for interactions (Cox regression analysis) are shown. Level of significance: *** = P < 0.001; ** = P < 0.01; * = P <0.05; ns = non-significant).

Comparison between forests			**Comparison between inside forest vs open field**		
Factor	**Log-rank statistic**	**Wald statistic**	**Factor**	**Log-rank statistic**	**Wald statistic**
Forest	18.2 (***)		Site	0.02 (ns)	
Fencing	979.8 (***)		Fencing	1015 (***)	
Species	37.7 (***)		Species	53.8 (***)	
Forest* Fencing		360.0 (***)	Site* Fencing		455.0 (***)
Forest * Species		2.6 (ns)	Site * Species		14.1 (***)
Fencing * Species		153.6(***)	Fencing * Species		126.0 (***)

4. Discussion

Grazing might have a positive or negative effect on the successful regeneration of woody plants in tropical forests (McEvoy et al., 2006; Miller et al., 2006). In the studied church forests grazing clearly had a strong negative effect on germination, seedling survival and growth. In the fenced plots, more seeds germinated, seedling survival was higher and seedlings grew faster. We observed signs of browsing and trampling damage in almost all seedlings in the unfenced plots. Similar results were found in other cattle exclusion studies, where cattle was found to have killed seeds and established seedlings and/or pressed seeds deeply in the soil causing their death (Hulme and Borelli, 1999; Smit et al., 2006a, b). Studies in Ethiopian highlands showed that heavy grazing pressure significantly increased surface runoff and soil loss and reduced infiltrability of the soil which in turn undermine suitability of sites for germination (Mwendera and Mohamed Saleem, 1997). Jimenez et al. (2005) showed that no seedlings of endemic Dwarf Pine survived after being exposed to grazing and trampling for 3-4 years in Northeastern Mexico. Many studies confirm that livestock grazing negatively influence survival and height-growth of most tree species, and that the effects are strongly related to grazing intensity (Teich et al., 2005; Opperman and Merenlender, 2000; Ramirez-Marcial et al., 2001; McEvoy et al., 2006).

Along the gradient from wet to dry areas in the tropics, we expect tree regeneration to change from being limited by light to being limited by moisture (van Bloem et al., 2004; Saw, 2004). In our study, seed germination was higher inside the forest than in the adjacent open area for all species. This most probably is attributed to differences in micro-environmental conditions, though we did not measure micro-environmental variables. Other studies showed that in the open field the solar insolation is high, resulting in high temperatures, low relative humidity, and higher irradiance, conditions that generally might cause death of seeds by desiccation (Pedraza and Williams-Linera 2003; Holl et al. 2000; Alvarez-Aquino et al., 2004). In fenced plots, however, for seedling survival and growth only minor differences were found between forest and open area suggesting that both light and moisture might limit regeneration at later phase (Cabin et al., 2002). This result suggests that church forests are between the extremes along the wet and dry gradient and that depending on local and temporal condition regeneration can be hampered by light and moisture stress. In unfenced plots, no seedlings survived until the end of the year neither in forest interior nor in open field, indicating that the grazing effect ultimately did not differ. The mortality rates, however, were higher in the open field.

The forests of Dengolt and Gelawdios have similar climatic conditions. The lack of interaction between forest and fencing showed that the forests had similar fencing effects on germination and seedling growth. However, seedling survival in the fenced plots in Dengolt was higher than in Gelawdios. This may be attributed to the fact that since Gelawdios forest is better developed and denser than Dengolt forest, it might harbor more rodents and other small wild mammals that could enter into the fences easily to consume the seedlings. During field visit we observed many rodents, perhaps from the genera Arvicanthis and Mastomy which are common in the area (Workneh, et al., 2004), porcupines, rabbits, and squirrels within the fenced plots. Moreover the availability of more light in the interior of Dengolt than Gelawdios could be another factor for the difference (Alvarez-Aquino et al., 2004)

The significant interaction between fencing and species on seed germination and seedling survival revealed that the magnitude of damage due to grazing can vary with species. This result confirms our hypothesis. The highest seed germination of *Prunus* may be attributed to its recalcitrant seeds, which germinate as quickly as possible before being predated or crushed (Pammenter and Berjak, 2000). *Olea* seedlings with higher survival rate in fenced plots did not survive in unfenced plots. In fenced plots in the open field they showed retarded growth, possibly indicating devastating effect of grazing/browsing by animals. It is possible that small goats, sheep, smaller wild mammals and rodents were still able to get inside the fenced plots and damaged the seedlings. On the other hand, in the unfenced plots, the mortality rate of *Ekebergia* was significantly lower than of the other species. *Ekebergia* may be more grazing tolerant or less attractive to herbivores compared to the other species. Some reports showed there are extractives from *Ekebergia capensis* like triterpenoids, tannins, limonoid which might have repellent effect on herbivores (Taylor, 1981; Nishiyama et al.,1996). From field observation we know that its leaves and also its fruits and seeds, have an unpleasant odor when crushed, possibly repelling predators. The shift of *Juniperus* from lower to higher survival and better growth in forest interior compared to outside the forest may indicate that this species does not perform good under forest shade (White, 1983; Hailu Sharew et al.,1996).

In conclusion, grazing strongly negatively effects seed germination, seedling survival and seedling growth, although species effects are different to some extent. This is even the case under incomplete animal exclusion circumstances. If the fences had been tighter, the results would have been even more pronounced. This negative grazing impact can be related to the lack of regeneration that was reported for a number of tree species (Linhart and Whelan, 1980; Teague, 1989; Jimenez et al., 2005). To achieve indigenous tree species regeneration in church forests controlling livestock pressure therefore is mandatory. Also, tree regeneration is more successful in forest interior than in open field, partly as a result of the higher grazing pressure outside the forest. As a result seeds naturally dispersed outside the forest will not have a chance to establish seedlings, grow and colonize the surroundings. This implies that livestock grazing is of paramount importance for both the internal sustainability of church forests and for restoration of the degraded surroundings. Possible solutions could be fencing against cattle grazing, seeding and planting in fenced areas, and planting browsing resistant species (Smit et al 2006) around the forest. Of course the awareness of local people in this respect is crucial. Creating awareness on the effect of grazing and on the benefit of exclosures for restoration could be a remedy before the damage to these forests by livestock become irreversible. The churches and their clergy members could play a pivotal role in this and promote long term sustainability of their forests.

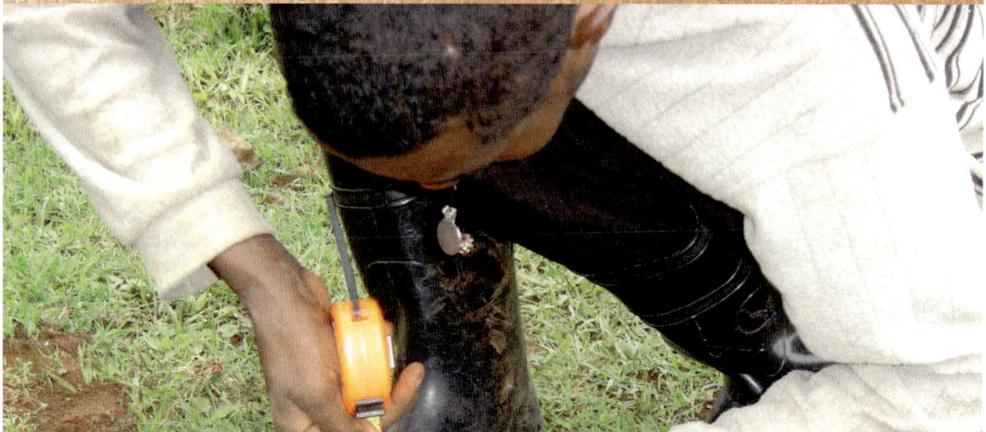

CHAPTER 6

TREE REGENERATION IN CHURCH FORESTS OF ETHIOPIA: effects of microsites, seed availability, litter and weed control

Alemayehu Wassie, Frank Sterck, Demel Teketay and Frans Bongers

Abstract

Tree regeneration in fragmented afromontane forests of northern Ethiopia is severely hampered. We explored how regeneration of trees in these remnant forests varies along the gradient from the forest interior to the edge and open fields, differs between canopy gaps and closed canopy sites inside the forest, and is affected by different management interventions, particularly seed sowing, litter removal and weeding. We focused on seed germination, seedling survival and growth of four indigenous tree species.

Seedling establishment was more successful inside the forest than outside. Seeds of *Juniperus, Olea* and *Ekebergia* germinated better in gaps while those of *Prunus* preferred a closed canopy. Seedling survival for three of the four species and seedling growth for all the species were highest in the outer edge of the forest. The combined effect of seed sowing and litter removal increased germination significantly for all the species. Weeding did not improve seedling growth of the species, and had even a negative effect on *Prunus*.

We conclude that though seedling establishment is higher inside the forest, seedlings preferred the outer edge of the forest for survival and growth. This may indicate that the immediate surrounding areas of the forest fragments need to be available to serve as buffer zones for the regeneration of tree species. In other words, given the surrounding land free from farming and grazing intervention, the seedlings of existing tree species can colonize and restore the forests on the degraded land. Within the forest, opening up of the canopy may be required to facilitate seed germination and seedling performance depending on the species. Sowing seeds combining with litter removal and slight soil scarification can be very useful. Weeding is not recommended since it did not have significant effects. These measures improve tree regeneration of these remnant forests.

1. Introduction

Worldwide, forests have been fragmented into small patches, and forest structure and regeneration have been influenced due to this fragmentation and habitat loss (Echeverria et al., 2006). The fragmentation results in small stands of forest bordered by open, often agricultural lands (Bustamante and Castor, 1998; Kollmann and Buschor, 2002; Fleury and Galetti, 2004). Consequently, such landscapes are dominated by strong environmental gradients from the forest interior to forest edges and open fields. Under the increasing pressure by the human population, via e.g. grazing activities or tree harvesting, natural regeneration may be hampered and, as a result, the persistence of the remnant forest patches and their indigenous species in many areas are threatened. This is particularly the case in the rapidly developing tropical areas (Cotler and Ortega-Larrocea, 2006). Here, we report on the regeneration of indigenous tree species in the fragmented Afromontane landscape in Ethiopia. While almost all dry afromontane forests have been converted to open agricultural lands (EFAP, 1994; Demel Teketay, 1996; Tesfaye Wubet et al., 2003), thousands of small isolated fragments remain around churches ("church forests") (Bingelli et al., 2003; Alemayehu Wassie et al.,2005a & b; Aerts et al., 2006a).

The regeneration of trees in church forests might be influenced by the environmental gradients that exists from the forest interior to forest edge to the open fields. The forest interior is characterized by a more stable humid and shaded microclimate compared to the agricultural open fields surrounding the forest (Jose et al., 1996; Gascon et al., 2000; Alvarez-Aquino et al., 2004). Closed forest patches may differ from open spots or gaps within the forest in a similar but less extreme environmental gradient (Denslow, 1980; Collins et al., 1985; Vitousek and Denslow, 1986; Bradshaw and Goldberg, 1989; Rincon and Huante, 1993). At both spatial scales, light may limit regeneration when moisture is not limiting (e.g. in the wet season), suggesting that regeneration may be facilitated in the more open sites. On the other hand, when soil moisture is mainly limiting (e.g. in dry season), regeneration may be facilitated in the more closed sites (Fetcher et al., 1985; Belsky et al., 1993; Gerhardt, 1996; Chen et al., 1999; McLaren and McDonald, 2003; Asbjornsen et al., 2004). The edges between exposed and closed areas may offer plants intermediate conditions, although there are also indications that they are subjected to stronger fluctuations in for example humidity and temperature compared to both forest and open field (Jose et al., 1996; Laurance and Bierregaard, 1997; Kapos, et al., 1997; López-Barrera and Newton, 2005). Forest edges, thus, produce a local environment that differs from both open areas and forest interiors (Chen et al., 1995). Other environmental variables may further complicate the possible outcome of regeneration, such as soil texture, litter layer, soil compaction, and the presence of potential competitors (Denslow, 1980). Here, we test whether open conditions facilitate or reduce the regeneration of indigenous tree species along the forest interior – edge – agricultural land gradients as well as between closed canopy and gaps in the forest interior.

Irrespective of exposure, the regeneration success can be hampered by seed availability, litter and weed competition. Successful establishment of seedlings could be hampered by low seed availability due to reduced seed rain and increased post-dispersal seed predation (Sanchez-Cordero and Martinez-Gallardo, 1998). An unfavorable microenvironment like a thick litter layer could also lead to either seed mortality or seed dormancy and deposition into the soil (as part of the soil seed bank) depending on the type of species. A thick litter layer decreases germination and seedling emergence through shading, biochemical effects and physical obstruction to the emergence of a seed's cotyledons and radicle (Facelli and Pickett, 1991; Molofsky and Augspurger, 1992; Ellsworth, et al., 2004). In contrast, a thick litter cover may protect large seeds from desiccation damage and predation (Myster and Pickett, 1993; Kitajima and Fenner, 2000). Although litter cover may have both negative and positive effects, a meta analysis of 32 published studies found that the overall effects of litter are negative on both seed germination and seedling establishment (Kitajima and Fenner, 2000). Weeds and herbs have also a significant role on tree seedling survival. They can have either facilitative or competitive effects. Weeds directly compete with seedlings for resources, a negative effect, but they can also diminish radiation loads, ameliorate disturbances or protect from herbivores. Different species have different tolerances to these interactions (Kitajima and Fenner, 2000; Benayas et al, 2005). In this study we explore how seed availability, litter and weeding may facilitate the regeneration of indigenous trees.

We aim to determine bottlenecks in the regeneration of indigenous tree species in the fragmented Ethiopian highland landscape with scattered church forest patches. These forests are preserved over centuries by the Ethiopian Orthodox Tewahido Church and are considered sanctuaries for different plant and animal species. Recent inventories, however, show that populations of the tree species in the church forests are small and decreasing in extent over time (Alemayehu Wassie, 2002; Bingelli et al., 2003; Alemayehu Wassie et al., 2005a). We set out to address the specific questions how regeneration of indigenous trees: (i) varies along the gradient from the forest interior to the edge and open fields; (ii) differs between canopy gaps and closed canopy sites inside the forest; and (iii) is affected by different management interventions, particularly seed sowing, litter removal and weeding.

2. Material and Methods

2.1. Study site

The study was conducted in South Gondar Administrative Zone (SGAZ), Amhara National Regional State, in northern Ethiopia. Based on a large inventory of church forests in the Zone one forest, Gelawdios, in which species diversity and regeneration assessment study had been conducted previously was selected for the regeneration studies reported here (Alemayehu Wassie, 2002; Alemayehu Wassie et al., 2005a). Gelawdios forest is located in 'Dera Woreda (District)', and its first church was constructed around 1500 A.D. It is located at 11° 38′ N and 37° 48′ E at an altitude of 2549 m. The current forest area comprises 100 ha. The mean annual rainfall is 1216.3 mm (range 1103-1336), and average daily temperature is 17 °C. The main rainfall period is between June and September and a lesser amount of rain is also expected in March and April. Cambisols and Andosols are the predominant soil types in the area (SGDA, unpublished). Based on the survey carried out, the status of regeneration of the different trees in the forest was very low (Alemayehu Wassie, 2002; Alemayehu Wassie et al., 2005a).

2.2. Study species

To assess regeneration, we selected four tree species : *Juniperus procera* Endl., *Ekebergia capensis* Sparrm, *Prunus africana* (Hook. f.) Kalkm and *Olea europaea* L. subsp. *cuspidata* (Wall. ex DC.) Cifferri (hereafter referred to by their genus names and detail description is presented in Chapter 5, Table 1). We focused on three aspects of the life history cycle of these species, i.e. seed germination, seedling survival and growth. The selected species have high abundance, high socio-economic value and limited regeneration with no or very few individuals in the lowest diameter classes (Alemayehu Wassie, 2002).

2.3. Experimental design

2.3.1. Study plots

In the forest, six transect lines were located from the forest interior to the edge and open fields in different compass directions, each crossing the forest boundary. Along each transect line, study plots, measuring 10 x 10 m, were established at the forest interior, the inner edge (just inside the forest), the outer edge (just outside the forest) and open field at a distance of 25 m from the forest edge, which are jointly referred to as microsites hereafter. Edge is defined, here, as the location where there is abrupt change in vegetation cover from both the forest trees greater than 10 cm dbh and forest understory species to other land use types. The study plots in the forest interior were located perpendicular to and at a mean distance of 96 ± 8 m from the edge (Laurance and Bierregaard, 1997).

At both the interior and inner edge of forest, one plot each was established under the closed canopy and in the gap, i.e. in a nested plot design. Canopy openness was estimated using Spherical Crown Densiometer. The mean values of canopy closeness were 94 % ± 2 for the closed canopy and 8 % ± 1.5 for the gap. One plot each was also established in the outer edge and outside of the forest.

2.3.2. Seed germination experiment

Half of the plots was used for seed germination experiments and the other half was used for seedling survival and growth experiments. The half plots used for the seed germination experiments were further subdivided into four 1 x 1 m experimental units. Each experimental unit received one of the following factorial treatments randomly: (i) seed sowing and cultivation; (ii) seed sowing without cultivation; (iii) no seed sowing but cultivation; and (iv) neither seed sowing nor cultivation (control). This was done for each of the four study species. Therefore, the experiment involved six transects (replicates) and six half plots (two at forest interior, two at the inner forest edge, one at the outer edge and one outside of the forest) with 16 experimental units each for the four study species that received four treatments each. Cultivation refers, here, to removing the leaf litter until the decomposed organic soil layer appears and then lightly scraping the compacted topsoil. In the experimental units selected to receive the seed sowing treatments, twenty seeds of each species were sown along a straight line with 10 cm intervals between seeds and at a depth equivalent to their seed diameter (Hewitt and Kellman, 2004). The positions of seeds were marked with wooden stick picks to facilitate follow-up of the experiment. Emergence of seedlings above the soil was monitored, and the seedlings were counted and tagged every month for a period of one year.

2.3.3. Seedling survival and growth experiment

The half plots allocated for the seedling survival and growth experiments were further subdivided into two 1 x 1 m experimental units. In each experimental unit, 5 seedlings of the same species were planted. Each experimental unit received one of the two treatments randomly, i.e. (i) complete clearance of weeds through out the experiment period; and (ii) no weeding (control). This was done for each of the four species. Therefore, the experiment involved six transects (replicates), six half plots (two at forest interior, two at the inner forest edge, one at the outer edge and one outside of the forest) with eight experimental units each for the four study species that received two treatments each and 1440 seedlings (six transects x six half plots x eight experimental units x five seedlings = 1440 seedlings). Prior to the experiment, the seedlings were raised in a nearby nursery having similar climatic conditions and soil types as the forest. Two months before planting time, the seedlings were exposed to full sunlight for hardening. After planting (end of June 2005) the total initial plant heights were recorded (from ground up to main apex). The mean heights of the seedlings were 9.6 cm ± 4.3 for *Juniperus,* 8.8 cm ± 3.6 for *Ekebergia*, 12.1 cm ± 4.9 for *Prunus* and 17.5 cm ± 5.7 for *Olea*. The survival and height of the seedlings were recorded every month for a total period of twelve months, i.e. up to the next rainy season.

2.4. Data Analyses

2.4.1. Seed germination experiment

The data collected from the seed germination experiment were analyzed using GLM univariate analysis. Microsites, canopy openness nested in plots established in the interior and inner edge of forest (only inside forest), treatments and species were used as independent factors while the number of germinated seeds was the respondent variable in a univariate analysis. Transects were used as random factor. The number of germinated seeds was log-transformed to fulfill the assumptions of parametric test, i.e. normality and homogeneity of variances. As there were zero values one was constantly added to each data set before transformation. To determine the significant differences among the effects of microsites, treatments and species, Tukey HSD *post hoc* tests were used (Zar, 1984).

2.4.2. Seedling survival and growth experiments

The seedling survival over time was analyzed using a time to event approach, which measures the time to an event for each seedling (Altman and Bland, 1998). In this study, the critical event was death of seedlings. The model used for the analysis was Kaplan-Meier Survival Analyses (Bland and Altman 1998). Equality of survival distributions for factors, i.e. microsites, canopy openness, treatments and species was tested using log-rank test (Kleinbaum and Klein, 2005). Cox-regression of survival analysis was used to analyse the interactions among factors explicitly (Kleinbaum and Klein, 2005).

The change in growth over time was assessed by calculating relative height growth rate (RGR) over the whole year (0-12 months), which expresses growth in terms of a rate of increase in size per unit of initial size (Evans, 1972; Hunt,1982). The formula used to calculate RGR was: RGR= $(\log_e H_n - \log_e H_o)/(t_n - t_o)$, where H_n- height measured at specified time; H_o - initial height measurement; $(t_n - t_o)$ - the unit time differences between specified periods. The average of RGR calculated for each seedlings was taken for the analysis. RGR was used since it accounts for differences in initial seedling height and depicts seedling growth potential better (Land and Rieske, 2006). The distribution of RGR met the requirements of parametric test, and microsite, canopy openness (nested inside the forest), treatment and species were incorporated as factors in a univariate analysis of variance. All statistical analyses were done using SPSS version 12.0.1 (Field, 2005).

3. Results

3.1. Seed germination

Microsite, treatments and species affected the mean number of germinated seeds significantly ($P < 0.05$) while canopy openness had no effect (Table 1). Mean number of germinated seeds was lowest outside the forest (2.8 ± 0.4), highest in the forest interior (7.2 ± 0.7) and intermediate in the inner (6.7 ± 0.6) and outer (4.2 ± 0.6) forest edges. The mean number of germinated seeds was always lowest for *Juniperus* (1.0 ± 0.1), highest for *Prunus* (10.7 ± 0.9) and intermediate for *Ekebergia* (7.2 ± 0.7) and *Olea* (4.3 ± 0.5). For all the species, the combined effect of seed sowing and cultivation treatments on the number of germinated seeds was always significantly higher (10.1 ± 0.5) followed by the separate treatment effects of seed sowing (5.5 ± 0.5) and cultivation (4.9 ± 0.8). The mean number of germinated seeds in the control plots was the lowest (2.7 ± 0.5) (Fig. 1). There were significant interaction effects ($P < 0.05$) between microsite and species, canopy openness and species as well as treatment and species on the germination of seeds that suggests distinct species specific response to these three factors. But, no significant interaction effects were found between microsite and treatment nor between canopy openness and treatment (Fig. 1 & 2, Table 1). *Juniperus, Ekebergia* and *Prunus* had higher number of germinated seeds in the interior than outside the forest and exhibited intermediate number of germinated seeds in the forest edges. Seeds of *Olea* did not respond significantly to microsite (Fig. 1). These differences among species are also reflected by the significant microsite-species interaction effect (Table 1).

Both species showed higher responses to the combined treatments of seed sowing and cultivation. However, *Juniperus* and *Olea* did not respond to the cultivation treatment alone. Inside the forest, seeds of *Juniperus* did not germinate at all while very few seeds of *Olea* germinated in the cultivated and control plots. Outside the forest, however, relatively more seeds of *Juniperus* and *Olea* germinated in the cultivated and control plots (Fig. 1). For *Ekebergia* and *Prunus,* only cultivation resulted in more germinated seeds than sowing seeds inside the forest while outside the forest, sowing seeds was more important for seed germination than cultivation alone (Fig. 1).

Within the forest, the significant interaction effect for canopy openness and species (Table 1) suggests that canopy openness effect also differed between species (Fig. 2). Higher number of *Ekebergia* seeds germinated in gaps, and seeds of *Prunus* germinated more under the closed canopy. Seeds of both *Juniperus* and *Olea* did not respond significantly to canopy openness (Fig 2).

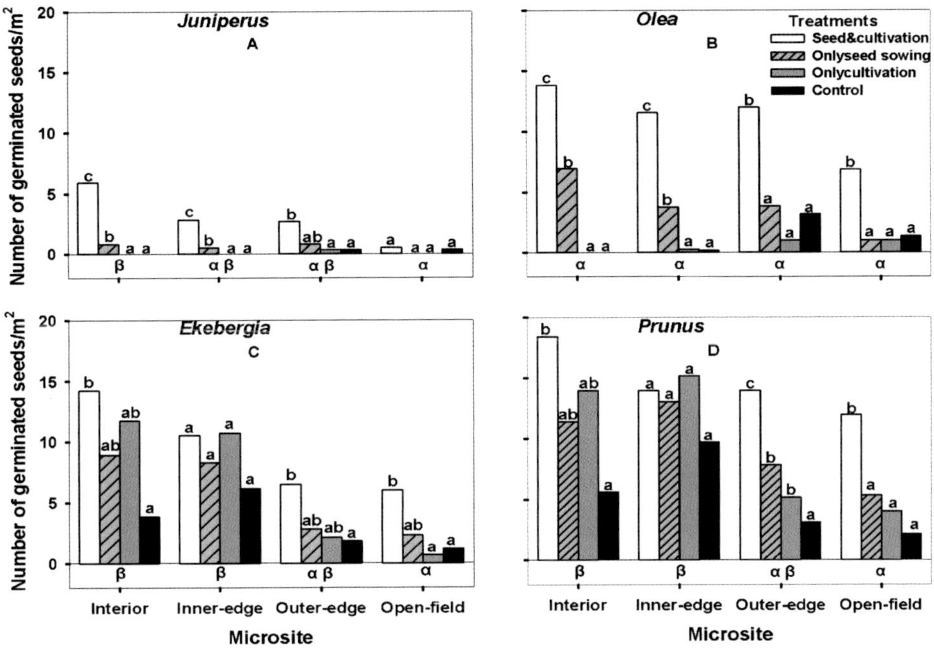

Figure 1. Effects of microsite, treatment and species on the number of germinated seeds for the four study species. Bars show mean values, capital letters differences between species, greek letters differences among microsites, small letters differences among treatments and different letters significant differences (P < 0.05).

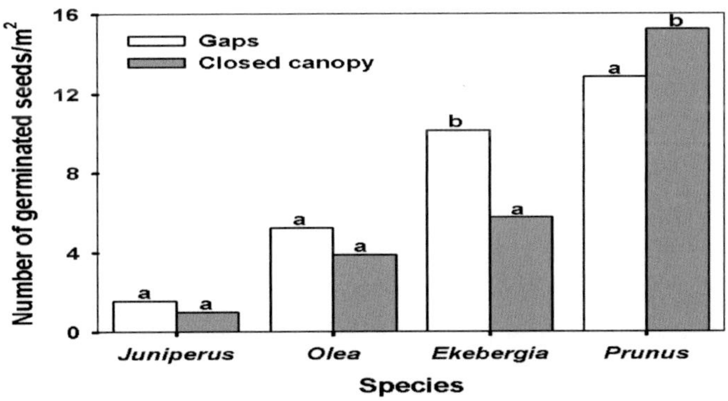

Figure 2. Effects of canopy openness and species on the number of germinated seeds for the four study species. Bars show mean values. small letters differences between gap and closed canopy plots for each species, different letters significant differences (P < 0.05).

3.2. Seedling survival

All factors analyzed had significant effects on the survival of seedlings of the study species (Kaplan-Meier Survival Analysis, Table 1). Seedling survival differed between microsites but did not follow a pattern along the forest interior to the open field gradient. Also, canopy openness and treatment had different effects on the survival of species. The trend were however complicated by the significant interaction effects ($P < 0.05$) between species and microsite, species and canopy openness as well as species and treatments (Table 1, Fig. 3 & 4).

With respect to microsite, seedling survival was lowest at the inner forest edge (36.3 %), highest at the outer forest edge (48.8 %), intermediate in the forest interior (44.8 %) and outside of the forest (41.3 %). The survival of *Olea* (83 %) was significantly higher ($P < 0.05$) than the other three species (23 - 31 %), which did not show significant differences in the survival of their seedlings (Fig. 3). Survival of *Juniperus* seedlings was higher in the outer forest edge (53 %), least in the inner forest edge (17 %) and intermediate in the forest interior (32 %) and outside the forest (26 %). In general, survival of *Juniperus* seedlings was higher outside than inside the forest (Fig. 4). On the contrary, more seedlings of *Prunus* survived in the forest interior (29 %) and inner forest edge (25 %) than in the outer forest edge (16 %) and outside the forest (15 %). Seedlings of *Ekebergia* survived more in the plots outside the forest and outer forest edge for six to seven months after which no significant difference in seedling survival was observed in the forest interior to open field gradient (25-38 %) (Fig. 3). Seedlings of *Olea* also showed the same survival trends as those of *Ekebergia* although the survival was higher in all cases. Death of seedlings of *Ekebergia* and *Olea* was low up to four to six months after planting with no differences among microsites (Fig. 3).

It is interesting to note that the control plots exhibited higher seedling survival than other plots, which received the weeding treatment. At the species level weeding resulted in a significant negative effect ($P < 0.05$) only for *Prunus*.

Seedling survival was higher in the gap (47.3 %) than in the closed canopy (33.7 %). Canopy openness did not significantly affect seedling survival of the three species for four to six months after planting while for *Ekebergia* a positive effect of gaps started so early. Later on, however, seedling survival in gap was higher than under closed canopy. The difference was significant ($P < 0.05$) for *Ekebergia* and *Olea* but not for *Juniperus* and *Prunus* (Fig. 4).

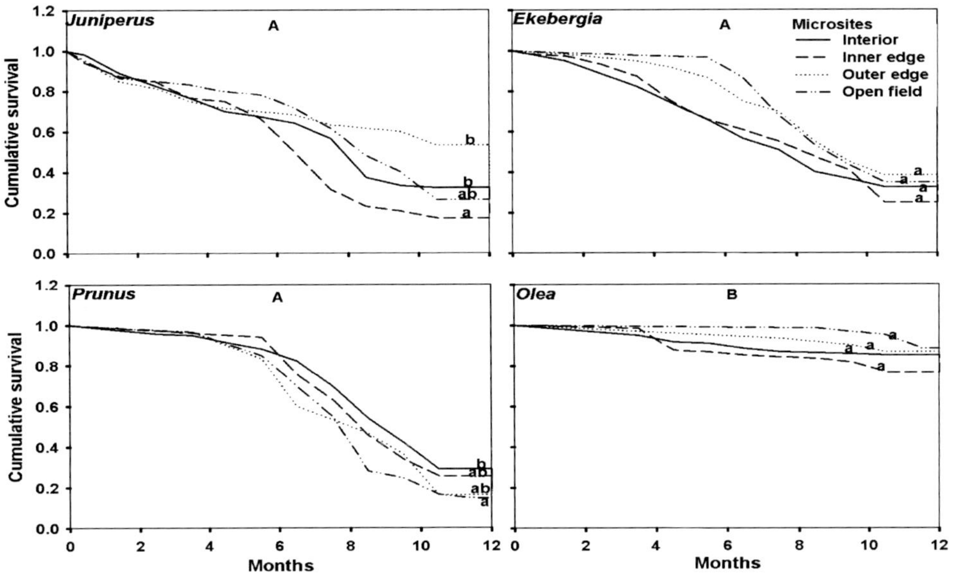

Figure 3. Seedling survival curves for microsite and species [capital letters show differences in seedling survival among species, small letters differences among microsites and different letters significant differences ($P < 0.05$)].

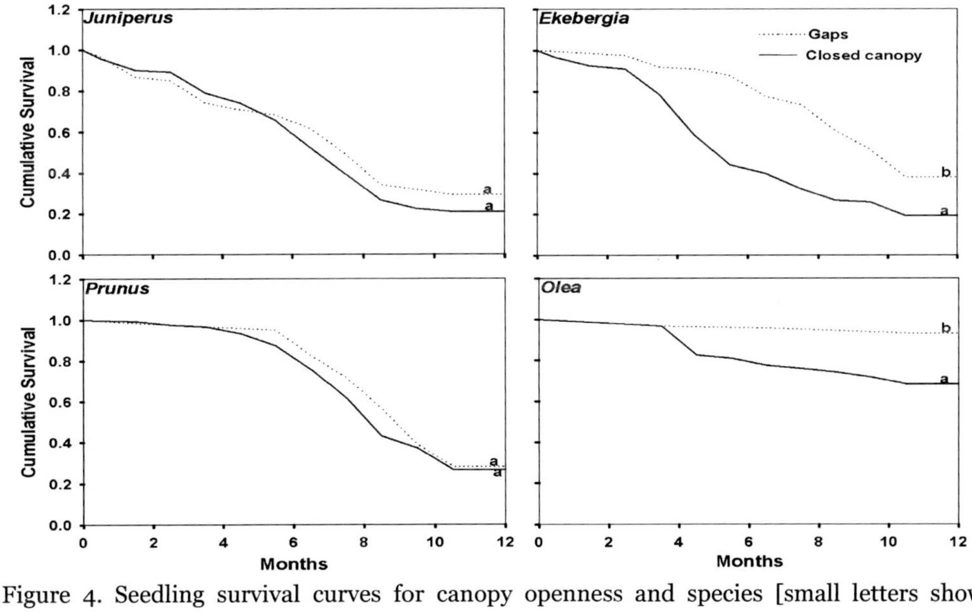

Figure 4. Seedling survival curves for canopy openness and species [small letters show differences in seedling survival between gap and closed canopy plots, and different letters significant differences ($P < 0.05$)].

3.3. Seedling growth

There were significant main effects (P < 0.05) of microsite, canopy openness and species on the RGRs of seedlings while treatment had no effect (Table 1). RGR of seedlings was lowest outside the forest (2.3 %/month), highest at the outer forest edge (5.7) and intermediate in the forest interior (5.1) and inner forest edge (3.9) (Fig. 5). Seedlings grew better in gaps (6.6 %/month) than under closed canopy (2.4 %/month) (Fig. 6). The lowest RGR was recorded for *Olea* (0.009 %/month), the highest for *Juniperus* (7.0) and the intermediate for *Ekebergia* (5.6) and *Prunus* (4.6).

There were significant interaction effects (P < 0.05) between species and microsite as well as between species and canopy openness while no interaction effects were found between microsite and treatment, canopy openness and treatment or species and treatment (Fig. 5 & 6, Table 1). Seedlings of *Juniperus* grew better in the forest interior than outside of the forest. *Ekebergia* and *Prunus* exhibited better growth in the outer forest edge though the RGR of seedlings was significant (P < 0.05) only for *Ekebergia.* In general, seedlings of *Olea* grew only inside the forest and experienced retarded growth outside the forest. Except for *Olea, seedlings of* the other species grew significantly higher (P < 0.05) in gaps. Seedlings of *Olea* showed no growth differences in gaps and the closed canopy.

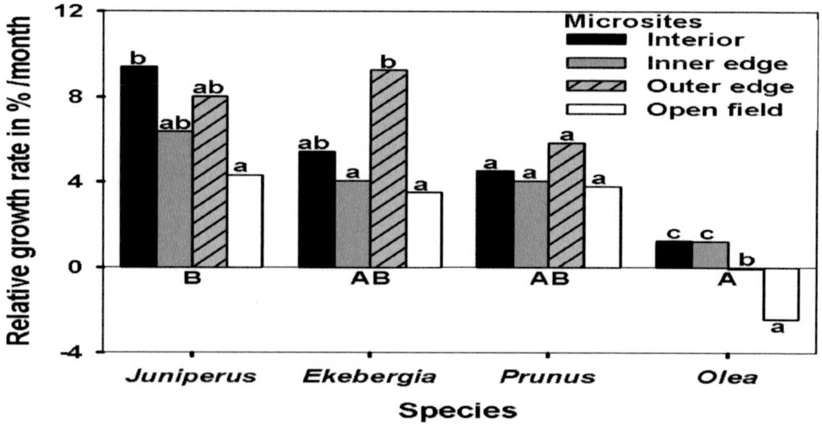

Figure 5. Effects of microsite and species on the relative height growth rate (RGR) of seedlings for the four study species. Bars show mean values (RGR/month), capital letters differences in seedling growth between species, small letters differences among microsites for each species and different letters significance differences (P < 0.05).

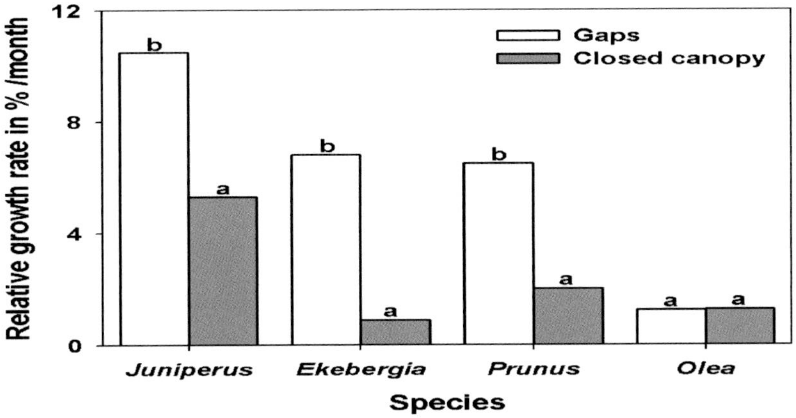

Figure 6. Effects of canopy openness and species on the relative height growth rate (RGR) of seedlings of the four study species. Bars show mean values (RGR/month), small letters show differences in seedling growth between gaps and closed canopy plots and different letters significance differences (P < 0.05).

Table 1. Effects of microsite, canopy openness, treatment and species with their interactions on number of germinated seeds, relative growth and survival of seedlings of the four study species. ANOVA results for germination and RGR with F and P values while for seedling survival Log rank test statistic for main effects (Kaplan-Meier survival analysis) and Wald statistic for interactions (Cox regression analysis) are shown; level of significance: ** = P < 0.001; * = P < 0.05; ns = non-significant).

Factors	Dependent variables					
	Number of germinated seeds		RGR (0-12) %/month		Survival of seedlings	
	Statistic	P	Statistic	P	Statistic	P
Microsite	15	**	8.6	**	13.6	*
Canopy openness	1.6	ns	20	**	29.7	**
Treatment	106	**	1.0	Ns	5.2	*
Species	15	**	60	**	274	**
Microsite * Treatment	1.6	ns	2.2	Ns	3.5	ns
Canopy openness * Treatment	0.67	ns	2.3	Ns	3.7	ns
Microsite * Species	1.9	*	3.8	**	69.8	**
Canopy openness * Species	10	**	4.2	**	34.2	**
Treatment * Species	4.6	**	0.76	Ns	8.9	*

4. Discussion

Effect of microsites

In dry tropical forests, light and soil moisture jointly regulate tree regeneration (Rincon and Huante, 1993; Ray and Brown, 1995; Khurana and Singh, 2000; McLaren and McDonald, 2003). The interplay effect of light and moisture depends on dry or wet seasons (Gerhardt, 1996), microsites, e.g. inside or outside forest, and gaps or closed canopy (Lieberman and Li, 1992; Rincon and Huante, 1993; Hewitt and Kellman, 2004) and/or the types of species involved (Khurana and Singh, 2000).

Except for *Olea* for which the difference was non significant, seedling establishment was more successful inside the forest than outside. Germination and early seedling establishment in the dry tropics are dependent mainly upon moisture availability (McLaren and McDonald, 2003). Some studies in tropical forests also showed that there is a soil moisture gradient from the forest interior and edges to outside forest areas (Williams-Linera, 1990; Brothers and Spingam, 1992; Camargo and Kapos, 1995; Jose et al., 1996), or more generally, inside to outside forests. Most likely, this moisture gradient might have brought the variation in the seed germination although we did not measure the moisture conditions in our study site. In a Caribbean dry forest, Ray and Brown (1995) found that for most species germination success was approximately equal in shaded and un-shaded sites.

The exposure effect between forest interior and exterior were not analogous to the exposure effect of gaps inside the forest. Inside the forest, seeds of *Juniperus, Olea* and *Ekebergia* germinated better in the gap than under the canopy shade though the difference was statistically significant only for *Ekebergia*. In the case of *Prunus*, more seeds germinate under the closed canopy than in gaps. Our results agree with those found by other authors. For instance, Demel Teketay and Granström (1997) and Negash Mamo et al. (2006) found that the overall germination performance of *Juniperus* seeds incubated under continuous light under laboratory conditions was significantly higher than those incubated in darkness. However, seeds collected from some populations germinated equally well in light and darkness (Negash Mamo et al., 2006). *Juniperus, Olea* and *Ekebergia* were favored by light for germination in the presence of better soil moisture inside the forest while *Prunus* remained persistently shade demanding. Outside the forest, where lower soil moisture is expected, the availability of full light did not result in a higher number of seed germination. This result suggests that germination is primarily influenced by moisture availability along the forest interior-open field gradient, and light within the forest.

Seedling survival and growth did not follow this trend. Seedling survival for three of the four species and seedling growth for all the species were highest in the outer edge of the forest. *Prunus* behaved as a shade tolerant species and survived better in the forest interior. This could be attributed to the fact that once the seedlings are established, these species demand more light and sites free from dominance by mature trees (Alvarez-Aquino et al., 2004). For instance, White (1983) designated *Juniperus* as a pioneer tree that does not regenerate under its own canopy. This was also confirmed by seedlings that responded to gaps for survival and growth though the level of significance depended on the species. Hailu Sharew et al. (1996) also found that relative height growth for *Juniperus* was higher in higher light conditions than deep shade under laboratory conditions that were designed to reflect the gaps and closed canopy conditions of afromontane forest.

Growth was not successful in the open field for all species. This probably was due to higher solar insolation, resulting in high temperatures, low relative humidity and higher irradiance (Alvarez-Aquino et al., 2004). The lowest survival rate at the inner forest edge may be attributed to the higher fluctuations of micro-environmental variable in edges (López-Barrera and Newton, 2005), to which seedlings have difficulties to adapt.

In all microsites, mortality of seedlings of the four species was low during the four to five months after planting but increased thereafter. This period corresponded with the beginning of the drier season in the study site. The differences between microsites and canopy openness also changed with time. For instance, seedling survival was higher outside than inside the forest for all the species up to four to five months after planting, but then the trend changed. This suggests the existence of a general tradeoff in dry forests that increases seedling growth and survival with increasing light during the wet period but greater mortality with increasing light during the dry period (Lieberman and Li, 1992; Rincon and Huante, 1993; Gerhardt, 1996; Demel Teketay, 1997; Cabin et al., 2002). As a result of the interplay effects of light and moisture, the outer forest edge might become the optimal place for seedling survival and growth. Within the forest, light in gaps favored survival and growth of seedlings both in dry and wet periods. Perhaps, the soil moisture in gaps inside the forest was not as depleted as that outside the forest in the dry period (Jose et al., 1996).

Effect of treatments

The combined effect of seed sowing and cultivation increased germination significantly for all the species. This implies that insufficient availability of seeds could be one of the constraints for regeneration in church forests. Seeds may not be available for germination in the forest due to reduced seed rain, pre- and post-dispersal seed predation or dormancy (Demel Teketay and Granström, 1997; Sanchez-Cordero and Martinez-Gallardo, 1998). Unavailability of seeds could also be due to litter layer on the forest floor, which could inhibit germination of the available seeds (Ellsworth et al., 2004). When we examine the impact of sowing seeds and cultivation separately, the former was more important than the latter for *Juniperus* and *Olea* indicating that seed availability is critical for these species. In agreement with our finding, Aerts at al. (2006c) reported that the humus layer did not have significant effect on seed germination of *Olea* in a drier shrub land. In another study, Hailu Sharew et al. (1997) and Eshetu Yirdaw and Leinonen (2002) reported that seeds of *Juniperus* were inhibited from germinating by the forest litter while ground preparation significantly increased their germination. However, *Prunus* and *Ekebergia* exhibited better germination of seeds in response to cultivation than seed sowing. This could be attributed to the possibility that there was few inputs from the seed rain and the critical problems remained was the hindrance effect of litter. Inside the forest, where more seeds are expected from the seed rain, cultivation was more important while seed sowing was more important for these two species outside the forest. It is interesting to note that no seeds germinated in the cultivated and control plots inside the forest for *Juniperus* and *Olea*. This may indicate reduced seed rain, severe seed predation or absence of a persistent soil seed bank inside the forest. Results from an earlier study in the same forest (Alemayehu Wassie and Demel Teketay, 2006) showed that both species did not possess soil seed banks. As a result, the only possible source of seeds for the species is only the seed rain, which is highly dependant on the status of the standing population and success of the trees in producing seeds.

Weeding did not significantly improve seedling growth for all the species (Madsen, 1995) and affected their survival negatively. This was contrary to our expectation. The number of seedlings in the control plots was significantly higher than in the weeded plots for *Prunus* while no differences were found for the other three species. Cabin et al. (2002) also reported that weeding did not affect the regeneration of native species in a Hawaiian dry forest. The possible explanation for the higher number of seedlings of *Prunus* in the control plots could be attributed to the facilitative effect of weeds (Callaway, 1995; Brooker & Callaghan,

1998; Callaway et al., 2002; Kikvidze et al., 2006). These could be provision of shade, diminishing radiation loads on the ground reducing evaporation, amelioration of disturbance and/or protection from herbivores (Callaway et al., 2002; Benayas et al., 2005; Miller et al., 2006). Various studies showed that the facilitative effects increases in intensity or importance with the increase in stress of the abiotic environment (Bertness and Callaway, 1995; Pugnaire and Luque, 2001; Choler et al., 2001; Callaway et al., 2002; Brooker et al., 2005; Kikvidze et al., 2006). This can be linked to the abiotic environment in the study forest, which might have been so stressful for seedling survival and, thus, the growth of weeds with seedlings of *Prunus* resulted in facilitative rather than negative effects.

Variation in species responses

The pooled number of germinated seeds from all the plots was very low for *Juniperus* and *Olea* and highest for *Prunus*. Slow germination and relatively low germination percentages are typical for *Juniperus* due to a complex dormancy mechanism (Laurent and Chamshama, 1987; Demel Teketay and Granstörm, 1997; Eshetu Yirdaw and Leinonen, 2002). The seed dormancy explains the unreliable natural regeneration of this species (Hailu Sharew et al., 1996; Eshetu Yirdaw and Leinonen, 2002). Some studies recommended pre-sowing seed treatment, mainly mechanical scarification, to increase the germination success of *Olea* (Demel Teketay and Granström, 1997; Aerts et al., 2006c). Seeds of *Ekebergia* (Demel Teketay and Granström, 1997) and *Prunus* do not posses dormancy and, hence, germinate quickly and in large numbers.

Juniperus, *Ekebergia* and *Olea* survived better outside the forest while *Prunus* did better inside the forest. Survival of seedlings of *Juniperus* and *Prunus* under canopy shade was not significantly different from that in gaps while the survival of seedlings of *Ekebergia* and *Olea* was higher in gaps than under canopy shade. Our result was supported by studies carried out on three of our study species. Demel Teketay (1997) found more naturally regenerated seedlings of *Ekebergia* in gaps than under forest shade in a site having similar climatic conditions to our study site while *Juniperus* did not show any significant differences. Aerts et al. (2006c) found more seedlings of *Olea* regenerating naturally beneath canopies of pioneer shrubs than in open patches in drier parts of Ethiopia, but strong shade tends to lower the performance of seedlings. Similarly, Masresha Fetene and Yonas Feleke (2001) showed that *Olea* exhibited a reduced relative growth rate at low light regimes. Interestingly, the highest survival success of *Olea* compared with its low growth confirms the growth-survival tradeoff as observed for other plant species

(Poorter and Kitajima, 2007). Bacelar et al. (2006) found that *Olea* has efficient tradeoff between growth and survival under stressful environment by which the species resists extended drought periods (Chartzoulakis et al., 1999; Giorio et al., 1999; Morse et al., 2002; Bacelar et al., 2004). Perhaps, this might be the mechanism that enabled the species to survive in the most degraded and drought prone areas of Ethiopia.

Implication for regeneration of species in church forests

Seed germination increased along the gradient from open field to forest interior for the four studied species suggesting that moisture drives seed germination differences in church forests. The overall poor germination of some of the species could lead to the decline in and their local elimination from the forest. Though seedling establishment is higher inside the forest, seedlings grew more rapidly and survived better on the outer edge of the forest. This may indicate that seedlings can colonize and restore the forests on the degraded land immediately surrounding the church forests, given the surrounding land is free from farming and grazing intervention. Within the forest, local canopy openings facilitate seed germination and seedling performance depending on the species. The fact that all the four study species do not possess soil seed banks may stress the importance of seed availability. Low seed availability might be one of the bottlenecks for regeneration which suggests that actively sowing seeds combined with litter removal and slight soil scarification might facilitate germination. Weeding is not recommended since it did not show significant effects. Future research, focusing on measuring and quantifying the differences of microenvironment variables among microsites in church forests related to seed germination, and seedling survival and growth is recommended. This will certainly improve the possibilities for active restoration of these highly valued church forests.

CHAPTER 7

GENERAL DISCUSSION

1. Church forests in the landscape

In Northern Ethiopia, the forests around churches ("church forests") is most what is left of the original dry afromontane forest cover and is a result of the traditional conservation system of the Ethiopian Orthodox Tewahido Churches (EOTC) (Yeraswork Admassie, 1995; Alemayehu Wassie et al., 2005a). The most dominant species in the church forests are *Juniperus* and *Olea,* which were species mainly composing the predominating diverse vegetation of dry afromontane forests of Northern Ethiopia since 1400 to 1700 AD (Logan, 1946; Friis, 1992; Darbyshire et al. , 2003). Not all church forests, however, are the remnant of the original forest of the area. In those churches established very recently, since 1960's after the original forest had gone already, their forest is mostly rehabilitated vegetation type and plantations mainly of exotic species. Church elders and scholars witnessed that most of the forest area around churches were by far more extensive than currently existing (Alemayehu Wassie, 2002). The Christian population in Northern Ethiopia inhabits mostly the highlands within the range of 1500 to 3500 masl. EOTC are distributed following this settlement pattern.

Church forests possess diverse woody species and accommodate good wood stock compared to some of the biggest continuous forest in the central highlands of Ethiopia (Tamirat Bekele 1994; Demel Teketay and Tamirat Bekele, 1995). In our 28 church forests covering only 500 ha, a total of 168 woody species are found. Out of the 23 threatened indigenous tree species listed as national priority and commercially important species, 15 of them were found in these churches (Bekele Million and Berhanu Leykun, 2001). Moreover two indigenous tree species (*Juniperus procera* and *Prunus africana*) currently included in the IUCN red list of threatened species are common to abundant to church forests (IUCN, 2006). These forests accommodate many species represented by single individuals (rare species) and also many species found only in single plots (unique species). Moreover, our species area curve showed increasing trends suggesting that more species are expected for larger sample sizes. The fact that we only surveyed 28 churches out of 1404 churches available in the study area strongly suggests that more species will be found if more church forests are assessed. Church forests are thus important refuges for a wide diversity of woody species of the area. A possible explanation for the presence of many rare and unique species in our forests could be that the continuous reduction of forest area and the ongoing degradation and increased level of disturbance in the remaining forest patches. This makes that church forests should be high in the priority of conservation efforts in this part of Ethiopia.

Church forests, inbedded in a heavily degraded landscape matrix, are of a paramount importance for their role in the landscape ecology of the region as *in situ* and *ex situ* conservation sites of indigenous plants, which would ultimately serve as sources of seeds / propagules / germplasm for future development of forests (Alemayehu Wassie et al. 2005a&b). Plantation forestry in Ethiopia, started in 1894, is almost solely practiced with exotic species (Pohjonen and Pukkala, 1990). The reason for the popularity of the exotic species is that they are fast-growing pioneers, easy to plant and care for compared to the lesser-known indigenous trees. But relying on a few exotic species increases the ecological risks of plantations. Increasing the share of indigenous tree species is becoming critically important (Pohjonen and Pukkala, 1992). The absence of natural forests that can serve as a seed source for the plantation of the native woody species diversity is a major limiting factor particularly in the degraded central and northern highlands of Ethiopia. However, the small isolated patches of more or less natural forest stands occurring around the numerous churches can be utilized as sources of diasporas (Eshetu Yirdaw, 2002). The church forests can also serve as a reference ecosystem for restoration efforts of the degraded landscape. They can provide knowledge and point of orientation to determine habitat of individual species and an association or assemblages of species and its synecology for communities (Alemayehu Wassie et al., 2005b; Bongers et al., 2006; Van Andel and Aronson, 2006).

We hypothesized that altitude, forest area, human influence via cattle interference and wood harvest are the major factors driving the variation in structural composition and species richness among these forests. We found that at higher altitude, forests are less diverse and that dominance by a single to few species is strong (Chapter 2). Similarly many studies in the tropics showed that species richness and diversity approximately linearly decrease with altitude (Stevens, 1992; Gentry, 1988; Aiba and Kitayama, 1999; Lacoul and Freedman, 2006). Our results showed that the altitudinal gradient is the main determinant of differences in species composition among these forests. Therefore their vast altitudinal distribution gives these forests the opportunity to hold most of the biodiversity resources of the area. This can be confirmed by the number of species we found in our 28 church forests (168) which is more than the expected number of species (125 woody species) listed in the study area by South Gondar Zonal Department of Agriculture (SGAZDA unpublished).

In contrast to our prediction, forest area did not influence the structure and species composition of church forests (chapter 2). This implies that although large sized forests are a necessary element of successful reproduction of species (Shafer, 1981;

Laurance and Yensen, 1991; Santos et al., 2007), small patches and appropriate matrix management could be useful complements for biodiversity conservation (Margules and Pressey, 2000; Fischer and Lindenmayer, 2001). On the other hand, the effects of forest area need time. As our result is based on spatial variation only, we may not completely disregard forest area effects on diversity and structure of forests in the long run (Debinski and Holt, 2000).

Although the local people generally respect the integrity of the church forests (Alemayehu Wassie, 2002), many of these forests are threatened by the increasing cattle population and increasing demand for wood products and farm lands. Human influence through livestock grazing and wood harvest had a strong effect on structural composition of the forest, but not on species richness and diversity except for understorey species richness. The interaction effect of altitude with human influence (particularly cattle interference) determined the species composition of the understorey of church forests (Chapter 2). Wood harvest also strongly affected the wood stock of these forests and aggravated degradation of the forests (Chapter 2).

The church has been using woods from these forests for church construction and mass services in a reasonably sustainable way for generations. However, the logging in some of the churches very recently is becoming much more intensive and in the interest of individuals instead of the church community. Undefined forest borders and irregular forest edges surrounded by individual land holdings have led to continuous but 'hidden' encroachments of agricultural fields into the forest. The conversion of the forest land to other land forms occurs in steps of only a few meters every year but given the small size of the forest, such changes might drastically reduce forest areas in the coming years and even lead to disappearance of the forests.

2. Regeneration in church forests

As a result of both human and environment induced factors, many of the church forests have only very few seedlings of woody species. In fact the understorey of the forests is completely dominated by shrubs and lianas (Chapter 2).

In one of our study forests (Gelawdios), we investigated the bottlenecks of regeneration at two different life phases: seed and seedling. In particular, the focus was on possible bottlenecks for the seed availability, seedling establishment, and seedling survival and growth (Fig. 1, Table 1). Major attention is given to four tree species that have high abundance and socioeconomic value but limited regeneration: *Ekebergia capensis* Sparrm, *Juniperus procera* Endl., *Olea europaea* L. subsp. *cuspidata* (Wall. ex DC.) Cifferri and *Prunus africana* (Hook. f.) Kalkm (hereafter referred to by their genus names).

Seed availability may be hampered due to limited seed rain, seed bank, and high seed predation. There are three possible sources of seeds for germination: local seed rain, soil seed bank and dispersal from outside. The availability of seeds is first determined by the amount of seed input from these sources. The annual seed input from the seed rain for the four species in the study site was for *Olea* = 4.7, *Ekebergia* = 5.4, *Juniperus* = 11.8 and *Prunus* = 22.9 seeds/m^2/yr and for the entire forest it was 327 seeds/m^2/yr (Alemayehu Wassie, unpublished data). The share of these four species from the entire annual seed rain in this forest was 13%. Soil seed bank analysis revealed that none of the four study species did possess a soil seed bank in Gelawdios (Chapter 3). Seed input from outside is unlikely as there are no forests left in most of Gelawdios forest surroundings. The available seed is further subjected to higher post-dispersal predation (*Ekebergia* = 98.3, Juniperus = 95.0, *Olea* = 89.2, *Prunus* = 100.0 %) (Chapter 4). Especially seeds of *Ekebergia* and *Prunus* suffered from high and rapid predation. Also their seeds suffer a high risk of microbial deterioration when buried (Chapter 4). Together these factors may, ultimately, limit the number of available seeds for regeneration. Seeds of *Juniperus* and *Olea* also suffer from high predation on the forest floor, but at least some of them remain viable until they get favorable conditions of which they germinate quickly (Chapter 4). In general, we found that seed availability was one of the limiting factors and mainly depends on seed rain for regeneration of *Ekebergia, Juniperus, Olea* and *Prunus* (Table 1, Chapter 6). Seed predation and short longevity of seeds in combination with a low seed input from the seed rain and seed bank may limit regeneration of these species. We expect this to be the case also for many other tree species in church forests.

We further looked to the factors influencing seedling establishment and seedling performance (Fig. 1). We hypothesized that the regeneration success of woody plants in church forests increases with forest size, and can be improved by reducing grazing, by microsite selection, and by management interventions particularly seed sowing, litter removal and weeding. In general, these predictions were confirmed qualitatively, but the factors differed in their quantitative effects on the regeneration success (Fig. 1). Additionally the effects also differed between species (Table 1).

Forest size

Although large sized forests are a necessary element of successful reproduction of many if not most tree species (Shafer, 1981; Laurance and Yensen, 1991; Santos et al., 2007), our results did not confirm that the density of seedlings increased with forest area (chapter 2, Fig. 1). Perhaps, the effects of forest area need time. As our result is based on spatial variation only, we may not completely disregard forest area effects on regeneration of forests in the long run (Debinski and Holt, 2000).

Livestock grazing

Our results confirmed that livestock grazing is the major factor limiting seedling establishment and seedling survival and growth in church forests. Almost none of the sown seeds were able to germinate in unfenced plots (Chapter 5: in unfenced plots germination was 4 and 5% in contrast to 57.5 and 63.3% in fenced plots in Dengolt and Gelawdios respectively and per species in unfenced plot *Ekebergia* = 4, Juniperus = 0, *Olea* = 3, *Prunus* = 13 % in Gelawdios, Table 1). Studies in Ethiopian highlands showed that heavy grazing pressure significantly increased surface runoff and soil loss and reduced infiltrability of the soil which in turn undermines suitability of sites for germination (Mwendera and Mohamed Saleem, 1997). The major challenge for seedlings survival and growth again is livestock grazing (Chapter 5). Seedling survival was higher (Chapter 5: *Ekebergia* = 48.3, Juniperus = 42.8, *Olea* = 90%, *Prunus* = 34.4 %) and seedlings grew faster (*Juniperus* = 7.4, Ekebergia = 6.6, *Prunus* = 8.5 *Olea* = 0.9 % RGR/month, Table 1) for all species in fenced plots than in unfenced plots (none survived) at the end of one year. We observed signs of browsing and trampling damage in almost all seedlings in the unfenced plots. In unfenced plots, no seedlings survived until the end of the year neither in forest interior nor in open field (chapter 5). As a result seeds naturally dispersed outside the forest will not have a chance to establish as seedlings, and then grow and colonize the surroundings. The difference between fencing and unfencing was clear even under incomplete animal exclusion circumstances, as our fencing was not able to completely exclude goats and sheep. Had the fences been tighter, the results would have been even more pronounced. Although in unfenced plots no seedlings survived until the end of the year neither in forest interior nor in open field, *Ekebergia* seedlings significantly survived longer than the other three species (chapter 5). This suggests that the effect of grazing on seedling survival and growth depends on the species. In general controlling livestock grazing is of paramount importance for both the internal regeneration of church forests and for restoration of the degraded surroundings.

Microsite

Apart from grazing, microsite also influenced the seedling establishment and seedling survival and growth success. Along the gradient of forest interior to edge and open field, in general, seedling establishment was more successful inside the forest and in particular in the gaps within the forest. Cierjacks and coworkers (2007) also showed that seed germination was significantly higher in forest interior than in the edge and outside in Ecuadorian tropical forest. The exposure effect between forest interior and exterior was not analogous to the exposure effect of gaps inside the forest. This result suggests that germination is primarily influenced by moisture availability along the forest interior-open field gradient (McLaren and McDonald, 2003) and by light within the forest (Kitajima and Fenner, 2000). At species level (Table 1), except for *Olea,* which didn't show significant differences, seedling establishment was more successful inside the forest than outside. Inside the forest, seeds of *Juniperus, Olea* and *Ekebergia* germinated better in the gap than under the canopy shade. For *Prunus*, more seeds germinated under the closed canopy than in gaps. This result suggests that *Juniperus, Olea* and *Ekebergia* germinated more successfully at sufficient light on moderate soil moisture condition inside the forest, while *Prunus* remained persistently successful in shade (chapter 6). Fashing (2004) also reported that *Prunus* seeds germinate successfully in shaded conditions in west Kenya.

Though seedling establishment is higher inside the forest, seedlings grew more rapidly and survived better at the outer edge of the forest. Our result revealed that the negative effect of edge on regeneration was seen at the inner edge while the outer edge had positive effect on seedlings survival. As a result of the interplay effects of light and moisture, the outer forest edge might become the optimal place for seedling survival and growth. Within the forest, light in gaps favored survival and growth of seedlings both in dry and wet periods. Perhaps, the soil moisture in gaps inside the forest was not as depleted as that outside the forest in the dry period (Jose et al., 1996). This may indicate that seedlings can colonize and restore the forests on the degraded land from immediate surrounding areas of the church forests, given that the surrounding land is protected from grazing intervention and farming (chapter 6). At species level (Table 1), *Juniperus*, *Ekebergia* and *Olea* seedlings survived better in the outer edge of the forest while *Prunus* did better inside the forest. Survival of seedlings of *Ekebergia, Juniperus* and *Olea* was higher in gaps than under canopy shade while *Prunus* survived more inside the forest. From this trend of microsite preference we conclude that *Juniperus*, *Ekebergia* and *Olea* prefer light shade to full exposure (Demel Teketay, 1997; Masresha

Fetene and Yonas Feleke, 2001; Aerts et al., 2006b) while *Prunus,* is shade tolerant (Kigomo, 1987; Tsingalia, 1988; Kiama and Kiyiapi, 2000). In general, the quantitative effect of microsite differed among species.

Management interventions

Management intervention in the form of seed sowing and litter removal significantly increased the germination success for all species (Chapter 6). Experimental seed sowing led to a significantly higher seedling establishment (Chapter 6). Other studies also report similar results to our findings (Cierjacks et al., 2007). This implies that insufficient availability of seeds could be one of the constraints for regeneration in church forests. Seeds may not be available for germination in the forest due to reduced seed rain, pre- and post-dispersal seed predation, dormancy, or lack of a persistent soil seed bank (Demel Teketay and Granström, 1997; Sanchez-Cordero and Martinez-Gallardo, 1998). We also showed that a thick litter layer inhibits germination as more seeds germinated in cultivated plots than in control for all four species (Chapter 6). Especially the combination of seed sowing and litter removal increased germination much more than the separate treatments for all four species.

On the other hand, perhaps due to the facilitative effect of grasses and herbs, weeding did not have a significant positive effect on seedlings survival and growth (Chapter 6). In contrast to our expectation, weeds even had a positive effect for *Prunus*. Some reports indicated that the balance between competition and facilitation shifts towards facilitation in stressful situations like moisture deficit and intense herbivore damage which are prevalent in church forests (Bertness and Callaway, 1995; Pugnaire and Luque, 2001; Choler et al., 2001; Callaway et al., 2002; Brooker et al., 2005; Kikvidze et al., 2006).

We conclude that the limited regeneration, with no or few individuals in the lowest diameter classes, observed for our four study species (Alemayehu Wassie, 2002) might be due to the factors associated to seed availability, grazing effect and microsite conditions. Low seed availability, aggravated by intense seed predation, and combined with a non-existing persistent soil seed bank is one of the bottlenecks for the regeneration of the four species. The few seeds able to pass this bottleneck are not able to establish seedlings successfully mainly due to grazing and a thick litter layer. Furthermore, grazing hampers the survival and growth of seedlings of the four species. Along the gradient from forest interior to edge and open field, seed sowing will be effective in the interior microsite, while for seedling transplant the outer forest edge is the preferable microsite. Within the forest, local canopy opening facilitates seed germination and seedling performance depending on the species. *Ekebergia, Juniperus and Olea* prefer microsites with light shade to full exposure while *Prunus* prefers more shady microsites for seed germination and seedling survival. The combined treatment of seed sowing and litter removal improves the seedling establishment of all species while herbs and grasses facilitate the survival and growth of *Prunus* seedlings (Table 1).

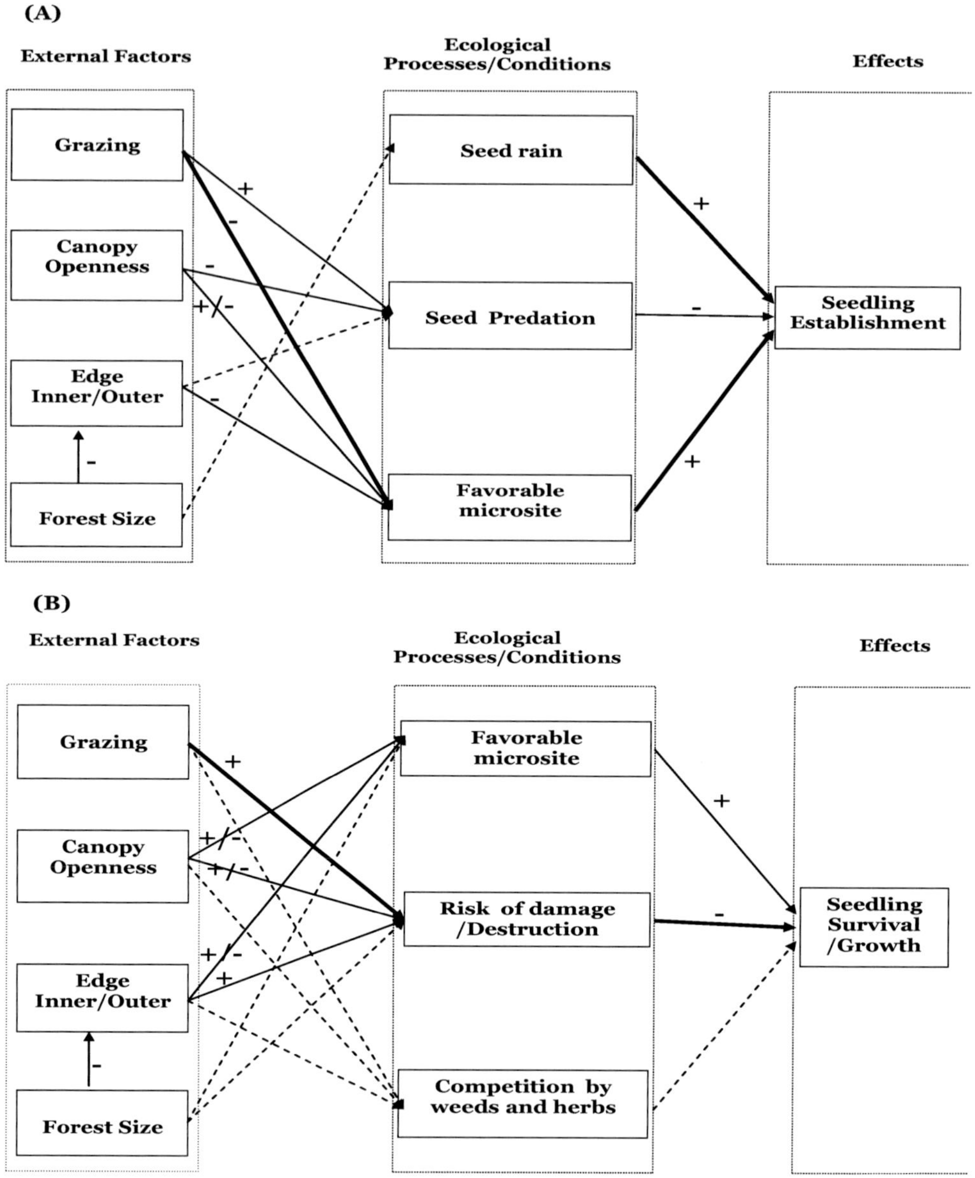

Figure1. Summary of factors that influenced (A) Seedling establishment (B) Seedling survival/growth of woody species in Ethiopian church forests. Bold line shows very strong effect, while broken line show that the expected effect was not supported by the analysis. The observed positive effects/relations are indicated by a + sign along the arrows, and the negative effects/relations by a – sign.

Table 1. Summary of factors affecting different life phases and possible management options for four tree species in Gelawdios forest. The bold figures show that they are significantly higher than the counterparts.

Life phase		Major bottlenecks for regeneration of the four species					Management options	Reference Chapter
		Factors	*Ekebergia*	*Juniperus*	*Olea*	*Prunus*		
Seed	availability	Seed rain (number/m^2/yr)	5.4	11.8	4.7	22.9	Reducing wood harvest of reproductive individuals; Seed sowing	* unpublished data
		Soil seed bank (number/ m^2)	0	0	0	0	Seed sowing	3
		Post dispersal predation (%)	98	95	89	100	Planting seedling; Seed sowing in enclosures; Providing alternative food sources for predators (e.g. crop seeds)	4
	Germination success	-With grazing -No grazing (%)	4 **81**	0 **22**	3 **57**	13 **93**	Livestock exclusion	5
		-Interior[1] -Inner edge[1] -Outer edge[1] -Open field[1] (%)	**71** 53 33 30	**30** 14 13 3	**69** 58 60 34	**92** 70 70 60	Sowing seeds in the interior and transplant seedlings to other microsite; Sowing seeds with mulching	6
		-Gap [1,2] -Closed canopy [1,2] (%)	**51** 29	**8** 5	**26** 20	64 **76**	Local canopy opening for *Ekebergia, Juniperus* and *Olea*	6
		-Control [1] -Sowing seeds & litter removal[1] (number/m^2)	3.3 **9.3**	0 **3**	1.3 **11**	5.1 **15**	Seed sowing combined with litter removal	6
Seedling	survival	-With grazing -No grazing (%)	0 **48**	0 **43**	0 **90**	0 **35**	Livestock exclusion	5
		-Interior[1] -Inner edge[1] -Outer edge[1] -Open field[1] (%)	33 25 **38** 35	33 18 **54** 27	85 77 88 88	**29** 26 17 15	Transplant seedlings of *Ekebergia* and *Juniperus* at the outer edge of the forest while for *Prunus* forest interior	6
		-Gap [1,2] -Closed canopy [1,2] (%)	**68** 27	38 29	**93** 28	19 21	Local canopy opening for *Ekebergia, Juniperus* and *Olea*	6
		- With weeds[1] - No weeds [1] (%)	29 33	32 28	83 83	**31** 17	Weeding not needed for *Prunus*	6
	Growth (RGR/month)	-With grazing -No grazing (%)	ns** 6.6	ns 7.4	ns 0.9	ns 8.5	Livestock exclusion	5
		-Interior [1] -Inner edge[1] -Outer edge[1] -Open field[1] (%)	5.4 4.1 **9.3** 3.5	**9.4** 6.4 **8.0** 4.3	**1.3** **1.2** -0.1 -4.7	4.5 4.1 5.8 3.8	Transplant seedlings of *Ekebergia* at the outer edge the forest	6
		-Gap [1,2] -Closed canopy [1,2] (%)	**6.8** 0.9	**10.5** 5.3	1.2 1.2	**6.5** 2.0	Local canopy opening for *Ekebergia, Juniperus* and *Prunus*	6
		- With weeds[1] - No weeds [1] (%)	5.7 5.0	7.8 7.9	1.4 0.04	**5** 3.6	Weeding not needed for *Prunus*	6

[1] plots without grazing; [2] plots with in the forest; * (Alemayehu Wassie, unpublished data); **ns refers to no survival at the end of the year

3. Management interventions and restoration opportunities

Ecological restoration efforts should both focus on maintaining forest species diversity and ecosystem services of the last remaining forests by creating conditions that facilitate regeneration of the various species within the forest as well as expanding forest area, interconnecting patches and creating new ones. These two efforts reinforce each other.

3.1. Facilitating regeneration in church forests

Based on our results (Fig 1. & Table 1) we here put forward several options to improve natural regeneration in church forests.

Seed phase

Reducing intensity of wood harvest will help to maintain adult trees in the forest which are crucial for seed supply and regular amelioration for regeneration (Table 1).

Sowing seeds might be considered as alternative to offset the seed limitation. Sowing seeds actively will be very important particularly for the four species (*Ekebergia, Juniperus, Olea and Prunus*) studied here and most probably for other species (e.g. rare species) for which sufficient seed rain is not expected. To be more effective, combining litter removal with slight soil scarification to loosen trampled soil is very useful. Sowing of seeds, should be combined with active reduction of predation and herbivory, e.g. in the form of enclosures (Fleury, and Galetti, 2006) and providing alternative food sources (Sullivan, 1978 & 1979) (Fig 1. & Table 1).

Reduce Grazing intensity to a level that does not destroy regeneration in church forests is mandatory and urgent (Fig 1. & Table 1). Both natural and subsidiary sowing seeds without livestock control are worthless. Possible solutions could be excluding cattle grazing in the entire forest for certain periods or seeding in fenced areas (Smit et al., 2006) around the forest.

Selecting/creating microsites: Along the gradient from forest interior to edge and open field, seed sowing will be effective in the interior microsite for better germination success . Within the forest, local opening up of the canopy may be required to facilitate both seed germination depending on the species.

Regeneration of species like *Ekebergia, Juniperus* and *Olea* can be facilitated by opening up intervention while shade tolerant species like *Prunus* prefers more shady micro environment. (Chapter 6) (Fig 1. & Table 1).

Seedling Phase

Transplanting seedlings can help to escape some of the germination bottlenecks. Especially in church forests where there is intense seed predation risk and germination is hampered by moisture deficit, seedling transplanting is an alternative till sufficient regeneration is achieved (Bergstern, 1985; Vallejo et al., 2006). Another advantage of transplanting is that nursery raised seedlings can be pre-conditioned to avoid transplant shock and to develop stress resistance so that their survival and growth after outplanting in the natural environment is improved (Landis et al., 1989). The major preconditioning technique includes manipulation of the watering regime and radiation environment (Vallejo et al., 2006). Especially for those rare species planting seedlings can ensure their future sustainability. Weeding is not recommended since it did not show significant effects. Even for species like *Prunus*, weeds like grass and herbs can have a facilitative effect (Fig 1. & Table 1).

Reduce Grazing intensity to a level that does not destroy regeneration in church forests is critical for seedling survival and growth (Fig 1. & Table 1). Both naturally rgenerated and transplanted seedling without livestock control are worthless. Possible solutions could be excluding cattle grazing in the entire forest for certain periods, seeding and planting in fenced areas, and planting browsing resistant species (Smit et al., 2006) around the forest.

Selecting/creating microsites: within the forest, local opening up of the canopy may be required to facilitate seedling performance depending on the species. Regeneration of species like *Ekebergia, Juniperus* and *Olea* can be facilitated by opening up intervention while shade tolerant species like *Prunus* prefers more shady micro environment. Along the gradient from forest interior to edge and open field, transplanting seedlings to the outer forest edges will be needed for better survival and growth (Chapter 6). The immediate surrounding areas of the church forests need to be available to serve as buffer zones for the regeneration of tree species. In other words, given the surrounding land free from farming and grazing intervention, the seedlings of existing tree species can colonize and restore the forests on the degraded land (Fig 1. & Table 1).

These interventions can facilitate seed germination, seedling survival and growth and ensure better regeneration in church forests which ultimately brings sustainability of these forests.

3.2. Interconnecting church forest

Forest remnants are parts of a landscape mosaic, and the presence of a given species in a patch may be a function not only of patch size and isolation, but also of the kind of neighboring habitat and of the species composition in the patches (Andrén, 1994; Wak et al. 1998). The effect of fragmentation on biodiversity is highly escalated when land use becomes so intense that corridors between patches disappear (Van Andel, 2006). Interconnecting church forests by vegetation corridors following natural terrain or stream lines, or reducing the distance between them by creating buffer areas around them and developing more patches in the landscapes (Simberloff, 1976; Hanski and Gyllenberg, 1993; Maschinski, 2006) are possible matrix management interventions. This will facilitate propagule and germplasm flow and ultimately sustain these forests and restore the surrounding landscape.

According to Krebs (2001) dispersal can follow three modes: (a) diffusion refers to a gradual movement of a population across a hospitable terrain through time; (b) jump dispersal refers to the movement of propagules over large distance and established successfully and; (c) secular dispersal refers to diffusion in evolutionary time. Accordingly the following strategies are practical options.

(i) Creating buffer areas and plantations around church forests enables to diffuse dispersal of species from the forest that later may inhabit the newly colonized area. Many places around churches can be assigned for this purpose. A household survey on four church localities concerning possible expanding of church forests, showed that most of the respondents (92.6%) were willing to plant seedlings around the existing church forests (Alemayehu Wassie, 2002). Direct seeding and planting of seedlings of keystone species can speed up restoration (Lamb et al., 1997; Sun et al., 1995; Camargo et al., 2002; Mulugeta Lemenih and Demel Teketay, 2005) especially when such species attract dispersing animals. Moreover, plantations may enhance the recruitment, establishment and succession of native woody species by functioning as foster ecosystems (Lugo, 1997; Eshetu Yirdaw, 2001; Feyera Sebeta et al., 2002). In this context the current trend of planting exotic species around most church forests, which mainly is for income generation, can play also a dual purpose in fostering restoration of native species.

(ii) Area enclosure promotion in the vicinity of church forests provides more safe sites for the jump dispersals (Harper, 1977). The mechanism of natural regeneration in area enclosures is based on vegetative recovery of buried plants and chance of seeds immigration from the outside (Teferra Mengistu, et al., 2005). It is very slow and the outcome is unpredictable. Especially in areas like northern Ethiopia where most of the top soil is removed and most woody species do not posses persistent soil seed bank (Demel Teketay and Granstorm, 1997; Alemayehu Wassie and Demel Teketay, 2006), there is little chance of recovery from soil seed bank. As a result recovery of tree species in enclosures depends mainly on seed dispersal from nearby forest remnants (Aerts et al., 2006b). Several government and non-government organizations have initiated enclosures programs aiming to restore the natural afromontane forest vegetation of Northern Ethiopia for the last 10 years (Asefa et al., 2003; Tefera Mengistu et al., 2005, Emiru Birihane, 2002). Strategically aligning these sites with respect to remnant forests is still lacking, however. Geographical distance and land features can affect dispersal movements. Most of the afromontane tree species exhibit limited long-distance seed dispersability (Demel Teketay and Granström, 1995). The larger the distance between enclosure site to remnant forests, the lesser chance for disperses to reach to enclosures (Aerts et al., 2006b). Dispersal by water (run off) needs specific alignment of remnants to enclosure (top to down) whereas for birds and mammals distance is most important. Enclosures situated along the down slope of remnant forest patches benefit from seed dispersal by run off (Aerts et al., 2006d). In general strategical alignment of enclosures that considers minimum possible distance to church forests will speed up restoration. This strategy benefits both enclosures by dictating them towards restoration, and church forests by ensuring sustainability as it ultimately reduces isolation and leads to increased connectivity (e.g. in Cunningham and Saigo, 1995; Falk et al., 2006).

4. Conclusion

In conclusion, in spite of what the Ethiopian Church has done to conserve significant proportions of the forests in Ethiopia, its forests do not receive the recognition and support they deserve. Because of this negligence, these forests are waiting for further degradation and possibly entire elimination comparable to the tragic fate of larger forests across the landscape in northern Ethiopia. Maintaining church forests is maintaining a large part of forest resources of northern Ethiopia.

Darbyshire (et al., 2003) showed that forests in northern Ethiopia have a potential to regenerate after as much as 1800 years (500 BC to 1300 AD) of anthropogenic clearance. However the newly regenerated forest type (*Juniperus* forest, with *Olea* and *Celtis)* is different from the earlier type (*Podocarpus-Juniperus* forest) (Darbyshire et al., 2003). Whether this change is a problem or not is left for discussion. However, such newly emerging ecosystems may help to strive to restore ecosystems that will be adaptive and resilient to local and global changes (Aronson and van Andel, 2006). The existing church forests provide great opportunities for restoration. They can serve as stepping stones to restore the surrounding degraded landscape. They can lead area enclosure programs to full scale restoration trajectory and in turn enclosures can ensure future sustainability of church forests. Therefore, preserving and restoring the remnant of indigenous forests, which are the home for native species in concrete ground before their entire elimination is, undoubtedly, an urgent task.

Ecological restoration efforts should both focus on maintaining forest species diversity and ecosystem services of the last remaining forests by creating conditions that facilitate regeneration of the various species within the forest as well as expanding forest area, interconnecting patches and creating new ones. The major management interventions that improves regeneration of tree species in church forests presented in priority are: reducing grazing intensity, reducing intensity of wood harvest, transplanting seedlings, sowing seeds combining with litter removal and selecting/creating microsites.

If management interventions in church forests, plantations around church forests and area enclosure programs are well integrated, restoration of the lost vegetation in northern Ethiopia should again be possible. This actually demands more integrated effort from different stakeholders. Continued support, law and policy enforcement from state, dedication from church clergy, and awareness creation to get trustworthy of the community are crucial.

REFERENCES

Abdella Gure, 2004. Seed-borne Fungi of the Afromontane Tree Species *Podocarpus falcatus* and *Prunus africana* in Ethiopia. Doctoral thesis: Swedish University of Agricultural Sciences Department of Forest Mycology and Pathology, Uppsala.

Abrahamson, W. G., 1980. Demography and vegetative reproduction. Pp 89-106. In Solbrig, O. T. (ed.). Demography and evolution in plant populations. University of California Press. Berkeley, California. USA.

Aerts, R., K. Van Overtveld, M. Haile, M. Hermy, J. Deckers and B. Muys, 2006a. Species composition and diversity of small Afromontane forest fragments in northern Ethiopia. Plant Ecology 187: 127-142.

Aerts, R., W. Maes, E. November, A. Negussie, M. Hermy and B. Muys, 2006b. Restoring dry Afromontane forest using bird and nurse plant effects: direct sowing of Olea europaea ssp. cuspidate seeds. Forest Ecology and Management 230: 23-31.

Aerts, R., E. November, I. Van der Borght, M. Behailu, M. Hermy and B. Muys, 2006c. Effects of pioneer shrubs on the recruitment of the fleshy-fruited tree Olea europaea ssp. Cuspidate in Afromontane savanna. Applied Vegetation Science 9: 117-126.

Aerts, R., W. Maes, E. November, M. Behailu , J. Poesen, J. Deckers, M. Hermy and B. Muys, 2006d. Surface runoff and seed trapping efficency of shrubs in a regenerating semiarid woodland in northern Ethiopia. Catena 65: 61-70.

Aiba, S. and K. Kitayama, 1999. Structure, composition and species diversity in an altitude-substrate matrix of rain forest tree communities on Mount Kinabalu, Borneo. Plant Ecology 140: 139–157.

Aizen, M.A., P. Feinsinger, 1994. Forest fragmentation, pollination, and plant reproduction in a chaco dry forest, Argentina. Ecology 75: 330–351.

Alemayehu Wassie, 2002. Opportunities and prospects of the Ethiopian Orthodox Tewahido Churches in conserving Forest resources: The case of churches in South Gondar, Northern Ethiopia. A master's degree thesis, Swedish University of Agricultural Science.

Alemayehu Wassie, Demel Teketay and N. Powell, 2005a. Church forests in North Gondar Administrative Zone, northern Ethiopia. Forests, Trees and Livelihoods 15: 349-374.

Alemayehu Wassie, Demel Teketay and N. Powell, 2005b. Church forests provide clues to restoring ecosystems in the degraded highlands of Northern Ethiopia. Ecological Restoration 23: 2.

Alemayehu Wassie and Demel Teketay, 2006. Soil seed banks in church forests of northern Ethiopia: Implications for the conservation of woody plants. Flora 201: 32–43.

Altman, D.G. and J.M. Bland, 1998. Statistics notes—time to event (survival) data. Br. Med. J. 317: 468-469.

Alvarez-Aquino, C., G. Williams-Linera and A.C. Newton, 2004. Experimental Native Tree Seedling Establishment for the Restoration of a Mexican Cloud Forest. Restoration Ecology 12: 412-418.

Amiaud, B. and B. Touzard, 2004. The relationships between soil seed bank, aboveground vegetation and disturbances in old embanked marshlands of Western France. Flora 199: 25-35.

Andrén, H., 1994. Effects of habitat fragmentation on birds and mammals in landscapes with different proportions of suitable habitat: a review. Oikos 71: 355-366.

Aronson, J. and J. Van Andel. Challenges for ecological theory. Pp223-233. In In Van Andel, J and J. Aronson (eds.), Restoration Ecology. Blackwell Publishing.

Asbjornsen, H., M.S. Ashton, D.J. Vogt and S. Palacios, 2004. Effects of habitat fragmentation on the buffering capacity of edge environments in a seasonally dry tropical oak forest ecosystem in Oaxaca, Mexico. Agriculture, Ecosystems and Environment 103: 481-495.

Asefa, D., G. Oba, R. B. Weladji and J. E. Colman, 2003. An assessment of restoration of biodiversity in degraded high mountain grazing lands in northern Ethiopia. Land Degradation and Development 14: 25–38.

Aymro, W. and J. Motovu, 1970. The Ethiopian Orthodox Church. The Ethiopian Orthodox Mission, Addis Ababa, Ethiopia.

Azene Bekele, 1993. Useful trees and shrubs for Ethiopia: Identification, Propagation and Management for Agricultural and Pastoral Communities. RSCU/SIDA.

Bacelar, E.A., C.M. Correia, J.M. Moutinho-Pereira, B.C. Gonc-alves, J.I. Lopes, J.M. Torres-Pereira, 2004. Sclerophylly and leaf anatomical traits of five field-grown olive cultivars growing under drought conditions. Tree Physiology 24: 233–239.

Bacelar, E.A., D.L. Santos, J.M. Moutinho-Pereira, B.C. Goncalves, H.F. Ferreira, C.M. Correia, 2006. Immediate responses and adaptive strategies of three olive cultivars under contrasting water availability regimes: changes on structure and chemical composition of foliage and oxidative damage. Plant Sci. 170: 596–605.

Barchuk, A.H., M.P. Diaz, F. Casanoves, M.G. Balzarini and U.O. Karlin, 1998. Experimental study on survival rates in two arboreal species from the Argentinean Dry Chaco. Forest Ecology and Management 103: 203-210.

Bazzaz, F. A. and S.T.A. Pickett, 1980. Physiological ecology of tropical succession: a comparative review. Annual review of Ecology and Systematics 11: 287-310.

Bekele Million and Berhanu Leykun, 2001 . State of Forest genetic Resources in Ethiopia. Sub-Regional Workshop FAO/IPGRI/ICRAF on the conservation, management, sustainable utilization and enhancement of forest genetic resources in Sahelian and North-Sudanian Africa (Ouagadougou, Burkina Faso, 22-24 September 1998). Forest Genetic Resources Working Papers, Working PaperForestry Department, FAO, Rome, Italy.

Belsky, A.J., S.M. Mwonga, R.G. Amundson, J.M. Duxbury and A.R. Ali, 1993. Comparative effects of isolated trees on their undercanopy environments in high- and low-rainfall savannas. Appl. Ecol. 30: 143-155.

Benayas, J. M. R., J. Navarro, T. Espigares, J. M. Nicolau and M. A. Zavala, 2005. Effects of artificial shading and weed mowing in reforestation of Mediterranean abandoned cropland with contrasting Quercus species. Forest Ecology and management 212: 302-314.

Bendz, M. 1986. Hillside closures in Wello. Ethiopian Red Cross society: Mission report. Vaxjo, Sweden.

Bengtsson, J., S.G Nilsson, A. Franc, P. Menozzi, 2000. Biodiversity, disturbances, ecosystems function and management of European forests. Forest ecology and management 132: 39–50.

Benitez-Malvido, J., 1998. Impact of forest fragmentation on seedling abundance in a tropical rain forest. Conservation Biology 12: 380-389.

Benitez-Malvido, J., 2001. Regeneration in tropical Rainforest Fragments. Pp 136-145. In Bierregaard Jr R.O., C. Gascon, T.E Lovejoy and R.C.G. Mesquita (eds.), Lessons from Amazonia: The ecology and conservation of fragmented forest. Yale University press, New Haven & London.

Benkman, C.W., 1995. The impact of tree squirrels (Tamiasciurus) on limber pine seed dispersal adaptations. Evolution 49: 585-592.

Bergstern, U., 1985. A study on the influence of seed predators at direct seeding of Pinus sylvestris L. Severiges Lanbruksuniversitet Rapporter13: 1-15.

Bertness, M.D. and R.M. Callaway, 1995. Positive interactions in communities. Trends Ecol. Evol. 9: 191-193.

Bingelli, P., Desalegn Desissa, J. Healey, M. Painton, J. Smith and Zewge Teklehaimanot, 2003. Conservation of Ethiopian Sacred Groves. European Tropical Forest Research Network. Newsletter 38: 37-38.

Blackmore, M. and P. M. Vitousek, 2000. Cattle Grazing, Forest Loss, and Fuel Loading in a Dry Forest Ecosystem at Pu'u Wa'aWa'a Ranch, Hawaii. Biotropica 32: 625-632.

Bland, J.M. and D.G., Altman, 1998. Statistics notes—survival probabilities (the Kaplan–Meier method). Br. Med. J. 317: 1572.

Bongers, F., A. Wassie, F. J. Sterck, T. Bekele and D. Teketay, 2006. Ecological restoration and church forests in northern Ethiopia. Journal of the Drylands 1: 35-45.

Bradshaw, L. and D.E. Goldberg, 1989. Resource levels in undisturbed vegetation and mole mounds in old fields. American Midland Naturalist 121: 176- 183.

Brokaw, N.V.L., 1982. The definition of treefall gap and its effect on measures of forest dynamics. Biotropica 14: 158-160.

Brooker, R. and T.V. Callaghan, 1998. The balance between positive and negative interactions and its relationship to environmental gradients: a model. Oikos 81: 196-207.

Brooker, R., Z. Kikvidze, F.I. Pugnaire, R.M. Callaway, P. Choler, C.J. Lortie and R. Michalet, 2005. Importance of importance. Oikos 109: 63-70.

Brothers, T.S. and A. Spingarn, 1992. Forest fragmentation and alien plant invasion of central Indiana old-growth forests. Conserv. Biol. 6: 91-100.

Bulinski, J. and C. McArthur, 1999. An experimental field study of the effects of mammalian herbivore damage on Eucalyptus nitens seedlings. Forest Ecology and Management 113: 241-249.

Bullock, J.M., 2000. Gaps and Seedling Colonization. Pp 375-395. In Fenner, M. eds. Seeds: The Ecology of Regeneration in Plant Communities. CABI Publishing, 2nd Edition.

Burkey, T.V., 1993. Edge effects in seed and egg predation at two Neotropical rainforest sites. Biological Conservation 66: 139–143.

Bustamante, R. O. and C. Castor, 1998. The decline of an endangered temperate ecosystem: the ruil (*Nothofagus alessandrii*) forest in central Chile. Biodiversity and Conservation 7: 1607-1626.

Cabin, R. J., S. G. Weller, D. H. Lorence, S. Cordell and L.J. Hadway, 2002. Effects of microsite, water, weeding, and direct seeding on the regeneration of native and alien species within a Hawaiian dry forest preserve. Biological Conservation 104: 181-190.

Cade, B.S, and J.D. Richards, 2005. User manual for BLOSSOM Statistical Software. Midcontinent Ecological Science Center U. S. Geological Survey.

Callaway, R.M., 1995. Positive interactions among plants. Bot. Rev. 61: 306-349.

Callaway, R.M., R.W. Brooker, P. Choler, Z. Kikvidze, C.J. Lortie, R. Michalet, L. Paolini, F.I. Pugnaire, B. Newingham, E.T. Aschehoug, C. Armas, D.

Kikodze, and B.J. Cook, 2002. Positive interactions among alpine plants increase with stress. Nature 417: 844-848.

Camargo, J.L.C. and V. Kapos, 1995. Complex edge effects on soil moisture and microclimate in central Amazonian forest. Tropical Ecology11: 205-221.

Camargo, J.L.C., I.D.K. Ferraz and A.M. Imakawa, 2002. Rehabilitation of degraded areas of Central Amazonia using direct sowing of forest tree seeds. Restoration Ecology 10: 636-644.

Carolina, A.H and A.S. Javier, 2001. The effect of introduced herbivores upon an endangered tree (*Beilschmiedia miersii*, Lauraceae). Biological Conservation 98: 69-76.

Chao, A., R.L. Chazdon, R.K. Colwell and T. Shen, 2005. A new statistical approach for assessing similarity of species composition with incidence and abundance data. Ecology Letters 8: 148–159.

Chartzoulakis, K., A. Patakas, A. Bosabalidis, 1999. Changes in water relations, photosynthesis and leaf anatomy induced by intermittent drought in two olive cultivars. Environ. Exp. Bot. 42: 113-120.

Chazdon, R. L. and N. Fetcher, 1984. Photosynthetic light environments in a lowland tropical rain forest in Costa Rica. The Journal of Ecology 72: 553-564.

Chen, J., J.F. Franklin, T.A. Spies, 1995. Growing-season microclimatic gradients from clearcut edges into old-growth Douglas-Fir forest. Ecol. Appl. 5: 74-86.

Chen, J., S.C. Saunders, T.R. Crow, R.J. Naiman, K.D. Brosofske, D. Mroz-Glenn, B.L. Brookshire, J.F. Franklin, 1999. Microclimate in forest ecosystem and landscape ecology: variations in local climate can be used to monitor and compare the effects of different management regimes. Bioscience 49: 288-297.

Choler, P., R. Michalet and R.M. Callaway, 2001. Facilitation and competition on gradients in alpine plant communities. Ecology 82: 3295-3308.

Cierjacks, A., J.E. Iglesias, K. Wesche and I. Hensen, 2007. Impact of sowing, canopy cover and litter on seedling dynamics of two *Polylepis* species at upper tree lines in Central Ecuador. Tropical Ecology 23: 309-319.

Collins, B.S., K.P. Dunne and S.T.A. Pickett, 1985. Response of forest herbs to canopy gaps. Pp. 218-234. In Pickett, S.T.A. and P.S. White, (eds.). The Ecology of Natural Disturbance and patchy Dynamics. Academic press, London.

Colwell, R. K. 2005. EstimateS: Statistical estimation of species richness and shared species from samples. Version 7.5. User's Guide and application published at: http://purl.oclc.org/estimates.

Connell, J.H., 1978. Diversity in tropical rain forests and coralreefs. Science 199: 1302–1310.

Cotler, H. and M.P. Ortega-Larrocea, 2006. Effects of land use on soil erosion in a tropical dry forest ecosystem, Chamela watershed, Mexico. Catena 65: 107-117.

Crawley, M.J., 1992. Seed predators and plant population dynamics. Pp 158–191. In Fenner, M. (Ed.), Seeds: The Ecology of Regeneration in Plant Communities. CAB International, Wallingford.

Crome, F.H.J., 1994.Tropical forest fragmentation: Some conceptual and methodological issues. Pp 61-77. In Moritz C. and J. Kikkawa (eds.), Conservation biology Australia and Oceania. Survey Beatty, Sydeny, Australia.

CSA, 2005. Central Statistical Agency of Ethiopia: National statistics.

Cufodontis, G., 1953–1972. Enumeratio Plantarum Athiopiae Spermatophyta, 2 vols. Jard. Bot. Etat., Bruxelles.

Cunningham, P.W. and .W.B. Sago, 1995. Environmental Science: a global concern, third edition. Wm. C. Brown Publishers.

Curtis, J.T. and R.P. McIntosh, 1950. The interrelations of certain analytic and synthetic phytosociological characters. Ecology31: 434-455.

Dale, V.H., S.C. Beyeler, B. Jackson, 2002. Understory vegetation indicators of anthropogenic disturbance in longleaf pine forests at Fort Benning, Georgia, USA. Ecol. Indic. 1: 155–170.

Darbyshire, I., H. Lamb and M. Umer, 2003. Forest clearance and regrowth in northern Ethiopia during the last 3000 years. The Holocene 13: 537-546.

De Souza, G. B., A. Blanco, and C.N. Gardenal, 2001. Genetic relationship among *Aedes aegypti* (Diptera: Culicidae) populations from Argentina using random amplified polymorphic DNA polymerase chain reaction markers. J. Med. Entomol. 38: 371-375.

De Steven, D., 1991. Experiments on mechanisms of tree establishment in old-field succession: seedling emergence. Ecology 72: 1066–1075.

De Villiers, A.J., M.W. Van Rooyen and G.K. Theron, 2003. Similarity between the soil seed bank and the standing vegetation in the Strandveld Succulent Karoo, South Africa. Land Degradat. Dev. 14: 527–540.

Debinski, D.M. and R.D. Holt, 2000. A Survey and Overview of Habitat Fragmentation Experiments.Conservation Biology 14: 342-355.

Demel Teketay, 1992. Human impact on a natural montane forest in southeastern Ethiopia. Mountain Research and Development 12: 393–400.

Demel Teketay, 1993. Problems associated with raising trees from seeds. The Ethiopian experience. In: Leith, H. and M. Lohmann (eds.), Restoration of Tropical Forest Ecosystems. Kluwer Academic.

Demel Teketay, 1996. Germination ecology of twelve indigenous and eight exotic multipurpose leguminous species from Ethiopia. Forest Ecology and Management 80: 209-223.

Demel Teketay, 1996. Seed ecology and regeneration in dry Afromontane forests of Ethiopia. Doctoral Thesis, Swedish University of Agricultural Sciences, Umea, Sweden.

Demel Teketay, 1997a. The impact of clearing and converting dry Afromontane forests into permanent arable land on the composition and density of soil seed banks. Acta Oecolog18: 557–573.

Demel Teketay, 1997b. Seedling populations and regeneration of woody species in dry Afromontane forests of Ethiopia. Forest Ecology and Management 98: 149–165.

Demel Teketay, 1998a. Soil seed bank at an abandoned Afromontane arable site. Feddes Repertorium 109: 161–174.

Demel Teketay, 1998b. The role of light and temperature in the germination of twenty herbaceous species from the highlands of Ethiopia. Flora 193: 411–425.

Demel Teketay, 1998c. The joint role of alternating temperatures and light quality in the germination of *Veronica anagallis-aquatica* and *V. javanica*. Trop. Ecol. 39: 179–184.

Demel Teketay, 1998d. Germination ecology of three endemic species (*Inula confertiflora, Hypericum quartinianum* and *Lobelia rhynchopetalum*) from Ethiopia. Tropical Ecology 39: 69–77.

Demel Teketay, 1998e. Environmental factors that control the germination of Solanum species from Ethiopia. Tropical Ecology 39: 79–87.

Demel Teketay, 2002. Germination responses of *Discopodium penninervium* to temperature and light. Flora 197: 76–80.

Demel Teketay and A. Granström, 1995. Soil seed banks in dry afromontane forests of Ethiopia. Vegetation Science 6: 777-786.

Demel Teketay and Tamirat Bekele, 1995. Floristic composition of Wof-Washa natural forest, Central Ethiopia: Implications for the conservation of biodiversity, Feddes Repertorium106: 127-147.

Demel Teketay and A. Granström, 1997a. Germination ecology of forest species from the highlands of Ethiopia. Tropical Ecology 13: 805-831.

Demel Teketay and A. Granström, 1997b. Seed viability of Afromontane Tree species in Forest soil. Tropical ecology13: 81-95

Denslow, J.S., 1980. Gap partitioning among tropical rainforest trees. Biotropica Supplement 12: 47-55.

Donoso, D.S., A.A. Grez and J.A. Simonetti, 2003. Effects of forest fragmentation on the granivory of differently sized seeds. Biological Conservation 115: 63-70.

Echeverria, C., D. Coomes, J. Salas, J.M. Rey-Benayas, A. Lara and A. Newton, 2006. Rapid deforestation and fragmentation of Chilean Temperate Forests. Biological Conservation 130: 481-494.

Edwards, S., Mesfin Tadesse, and I. Hedberg, 1995. Flora of Ethiopia and Eritrea: Vol. 2, Part 2. Addis Ababa, Ethiopia and Uppsala, Sweden.

Edwards, S., Mesfin Tadesse, Sebsebe Demissew and Hedberg, I. (eds.), 2000. Flora of Ethiopia and Eritrea: Vol. 2, Part 1. Addis Ababa, Ethiopia and Uppsala, Sweden.

Edwards, S., Sebsebe Demissew and I. Hedberg, (eds.), 1997. Flora of Ethiopia and Eritrea, vol. 6. The National Herbarium Addis Ababa University, Addis Ababa and Department of Systematic Botany, Uppsala University, Uppsala.

EFAP, 1994. Ethiopian forest action program, vol. II. The Challenge for Development. Ministry of Natural Resources Development and Environmental Protection. EFAP Secretariat, Addis Abeba, Ethiopia.

Ellsworth, J.W., R.A. Harrington and J.H. Fownes, 2004. Seedling emergence, growth, and allocation of oriental bittersweet: effect of seed input, seed bank, and forest floor litter. Forest Ecology and Management 190: 255-264.

Emiru Birhane, 2002. Actual and Potential Contributions of Enclosures to enhance Biodiversity in Dry lands of Eastern Tigray, with particular emphasis on Woody Plants. A master's degree thesis, Swedish University of Agricultural Science.

Ericksson, I., Demel Teketay and A. Granström, 2003. Response of plant communities to fire in an Acacia woodland and a dry Afromontane forest, southern Ethiopia. Forest Ecology and Management 177: 39-50.

Eshetu Yirdaw and K. Leinonen, 2002.Seed germination responses of four afromontane tree species to red/far-red ratio and temperature. Forest Ecology and Management 168: 53–61.

Eshetu Yirdaw, 2002. Restoration of the native woody-species diversity, using plantation species as foster trees, in the degraded highlands of Ethiopia. Doctoral thesis. Univ. Helsinki. Tropical Forestry Rep. 24.

Estrada, A. and R. Coates-Estrada, 1986. Frugivory in howling monkeys (Alouatta palliata) at Los Tuxlas, Mexico: dispersal and fate of seeds. Pp. 93–104. In Estrada, A., T.H. Fleming (eds.), Frugivores and Seed Dispersal. Dr. W. Junk Publ., Dordrecht.

Evans, G.C., 1972. The quantitative analysis of plant growth. Blackwell scientific publications, Oxford.

Facelli, J.M. and S.T.A. Pickett, 1991. Plant litter: its dynamics and effects on plant community structure. The Botanical Review 57: 1-33.

Fahrig, L., 2003. Effects of habitat fragmentation on Biodiversity. Annual. Rev. Ecol. Evol. Syst. 34: 487–515.

Falk, D.A., M.A., Palmer and J.B. Zedler (eds.), 2006. Foundations of Restoration Ecology. Society for Ecological Restoration International. Island Press, Washington.

FAO, 2001. Global Forest Resource Assessment 2000. FAO Forestry paper 140. FAO.

Feeny, P., 1976. Plant apparency and chemical defense. Recent Adv. Phytochem. 10: 1-40.

Fenner, M. and K. Kitajima, 1999. Seed and Seedling Ecology. Pp 589-611. In Pugnaire, F.I. and F. Valladares (eds.), Handbook of functional plant ecology. Marcel Dekker, Inc.

Fenner, M., 1985. Seed Ecology. Chapman & Hall, London-- New York.

Fenner, M., 1995. Ecology of seed banks. Pp. 507-528. In Kigel, J. and G. Galili (eds.), Seed development and germination. Marcel Dekker Inc., New York.

Fetcher, N., S.F. Oberbauer, B.R. Strain, 1985. Vegetation effects on microclimate in lowland tropical forest in Costa Rica. Int. J. Biometeorol. 29: 145-155.

Feyera S., Demel T. and B. Näslund, 2002. Native woody species regeneration in exotic tree plantations at Munessa-Shashemene Forest, southern Ethiopia. New Forests 24: 131-145.

Feyera Senbeta and Demel Teketay, 2001. Regeneration of indigenous woody species in the canopies of tree plantations in central Ethiopia. Tropical Ecology 42: 175–185.

Feyera Senbeta and Demel Teketay, 2002. Soil seed banks in plantations and adjacent dry Afromontane forests of central and southern Ethiopia. Tropical Ecology 43: 229–242.

Field, A., 2005. Discovering statistics using SPSS for windows. Sage publications. London.

Fischer, J., and D. B. Lindenmayer, 2002. Small patches can be valuable for biodiversity conservation: two case studies on birds in southeastern Australia. Biological conservation106: 129-136.

Fleischner, T.L., 1994. Ecological Costs of Livestock Grazing in Western North America. Conservation Biology 8: 629-644.

Fleury, M. and M. Galetti, 2004. Effects of microhabitat on palm seed predation in two forest fragments in southeast Brazil. Acta Oecologica 26: 179-184

Fleury, M. and M. Galetti, 2006. Forest fragment size and microhabitat effects on palm seed predation. Biological Conservation 131: 1-13.

Friis, I., 1992. Forests and forest trees of Northeast Tropical Africa. Kew Bull. Add. Ser. XV, 1–396.

Garin, I., A. Aldezabal, J. Herrero and A. García-Serrano, 2000. Understory foraging and habitat selection by sheep in a mixed Atlantic forest. Vegetation Science 11: 863-870.

Garwood, N.C., 1989. Tropical soil seed banks. Pp. 149-209. In Leck, M.A., V.T. Parker and R. L. Simpson (eds.). Ecology of soil seed banks. Academic Press, Inc., San Diego, California.

Gascon, C., G.B. Williamson and G.A.B., da Fonseca, 2000. Receding forest edges and vanishing reserves. Science 288: 1356-1358.

Gentry, A.H., 1988. Changes in Plant Community Diversity and Floristic Composition on Environmental and Geographical Gradients. Annals of the Missouri Botanical Garden 75: 1-34.

Gerhardt, K., 1996. Effects of root competition and canopy openness on survival and growth of tree seedlings in a tropical seasonal dry forest. Forest Ecology and Management 82: 33-48.

Gill, R.M.A. and V. Beardall, 2001. The impact of deer on woodlands: the effects of browsing and seed dispersal on vegetation structure and composition. Forestry 74: 209-218.

Gimaret-Carpentier, C., R. Pelissier, J.P. Pascal, and F. Houllier, 1998. Sampling strategies for the assessment of tree diversity. Vegetation Science 9: 161-172.

Giorio, P., G.Sorrentino, R. d'Andria, 1999. Stomatal behavior leaf water status and photosynthetic response in field-grown olive trees under water deficit. Environ. Exp. Bot. 42: 95-104.

González-Hernández, M.P. and F.J. Silva-Pando, 1996. Grazing effects of ungulates in a Galician oak forest. Forest Ecology and Management 88: 65-70.

Gotelli, N.J. and G.L. Entsminger, 2004. EcoSim: Null models software for ecology. Version 7. Acquired Intelligence Inc. & Kesey-Bear. Jericho, VT 05465. http://garyentsminger.com/ecosim/index.htm.

Granström, A., 1982. Seed banks in five boreal forest stands originating between 1810 and 1963. Can. J. Bot. 60: 1815–1821.

Greig, N., 1993. Regeneration Mode in Neotropical Piper: Habitat and Species Comparisons. Ecology74: 2125-2135.

Grime, J. P. and S.H. Hillier, 2000. The Contribution of Seedling Regeneration to the Structure and Dynamics of Plant Communities, Ecosystems and Larger Units of the Landscape. Pp 361-374. In: Fenner, M. (eds.) Seeds: The

Ecology of Regeneration in plant Communities. 2nd Edition. CABI publishing.

Grime, J.P., 1979. Plant Strategies and Vegetation Processes. Wiley, Chichester.

Grubb, P. J., 1977. Control of forest growth and distribution on wet tropical mountains: with special reference to mineral nutrition. Annual Review Ecol.Syst.8: 83-107.

Grytnes, J. A. and O. R. Vetaas, 2002. Species Richness and Altitude: A Comparison between Null Models and Interpolated Plant Species Richness along the Himalayan Altitudinal Gradient, Nepal. American naturalist march 159: 294-304.

Haig, A.R., U. Matthes and D.W. Larson, 2000. Effects of natural habitat fragmentation on the species richness, diversity, and composition of cliff vegetation. Can. J. Bot.78: 786-797.

Hailu Sharew, C.J. Legg and J. Grace, 1997. Effects of ground preparation and microenvironment on germination and natural regeneration of *Juniperus procera* and *Afrocarpus gracilior* in Ethiopia. Forest Ecology and Management 93: 215-225.

Hailu Sharew, J. Grace and C.J. Legg, 1996. Response of two Afromontane coniferous tree species to light and nutrient supply. Tree Physiology 16: 617-626.

Hanski, I. and M. Gyllenberg, 1993. Two general metapopulation models and the core-satellite hypothesis. American Naturalist 142: 17-41.

Harper, J.L., 1977. Population Biology of Plants. Academic Press, London.

Harrington, G.N., A. K. Irvine, F.H.J. Crome and L.A. Moore, 1997. Regeneration of large-Seeded Trees in Australian Rainforest Fragments: A Study of higher-Order Interactions. Pp 292-303. In Laurance, W. F. and R.O. Jr. Bierregaard (eds.) , Tropical Forest Remnants: Ecology, Management, and Conservation of Fragmented Communities. The University of Chicago Press, Chicago & London.

Harris, J.A. and R. van Diggelen, 2006. Ecological restoration as a project for global society. Pp 3-15 in Van Andel, J and J. Aronson (eds.), Restoration Ecology. Blackwell Publishing.

Hau, C.H., 1997. Tree seed predation on degraded hillsides in Hong Kong. Forestry Ecology and Management 99: 215-221.

Hedberg, I. And S. Edwards, (eds.), 1989. Flora of Ethiopia: Vol. 3. Addis Ababa and Asmara, Ethiopia and Uppsala, Sweden.

Hedberg, I. and S. Edwards, (eds.), 1995. Flora of Ethiopia and Eritrea, vol. 7. The National Herbarium, Addis Abeba University, Addis Abeba & Department of Systematic Botany, Uppsala University, Uppsala.

Hedberg, I., S. Edwards and N. Sileshi, (eds.), 2003. Flora of Ethiopia, vol. 4(1). The National Herbarium Addis Abeba University, Addis Abeba and Department of Systematic Botany, Uppsala University, Uppsala.

Heske E.J., J.H. Brown and Q. Guo, 1993. Effects of kangaroo rat exclusion on vegetation structure and plant species diversity in the Chihuahuan Desert. Oecologia 95: 520–524.

Hewitt, N. and M. Kellman, 2004. Factors influencing tree colonization in fragmented forests: an experimental study of introduced seeds and seedlings. Forest Ecology and Management 191: 39-59.

Hill, J.L. and P.J., Curran, 2001. Species composition in fragmented forests: conservation implications of changing forest area. Applied Geography 21: 157–174.

Hill, J. L. and P. J. Curran, 2005. Fragment shape and tree species composition in tropical forests: a landscape level investigation. African Journal of Ecology 43: 35–43.

Hill, S.C. and D.M. Morris, 1992. The Function of Seed Banks in Northern Forest Ecosystems. Queen's Printer for Ontario, Ontario.

Hobbs, R.J. and J.A. Harris, 2001. Restoration Ecology: Repairing the earth's ecosystems in the new Millennium. Restoration Ecology 9: 239-245.

Hoffmann L.A., E.F. Redente and L.C. McEwen, 1995. Effects of selective seed predation by rodents on shortgrass establishment. Ecol. Appl. 5: 200–208.

Holl K.D. and M.E. Lulow, 1997. Effects of species, habitat, and distance from edge on post-dispersal seed predation in a tropical rainforest. Biotropica 29: 459–468.

Holl, K. D., M.E. Loik, E.H.V. Lin, and I. A. Samuels, 2000. Tropical montane forest restoration in Costa Rica: overcoming barriers to dispersal and establishment. Restoration Ecology 8: 339-349.

Hopkins, M.S. and A.W. Graham, 1983. The species composition of soil seed banks beneath lowland tropical rain forests in north Queensland. Biotropica 15: 90–99.

Howe, H. F. and J. Smallwood, 1982. Ecology of seed dispersal. Annual Review of Ecology and Systematics 13: 201–228.

Hulme, P.E., 1996a. Herbivory, plant regeneration and species coexistence. Ecology84: 609–616.

Hulme, P. E., 1996b. Natural regeneration of yew (*Taxus baccata* L): microsite, seed or herbivore limitation? Ecology 84: 853–861.

Hulme, P.E and T. Borelli, 1999. Variability in post-dispersal seed predation in deciduous woodland: relative importance of location, seed species, burial and density. Plant Ecology145: 149-156.

Hulme, P.E., 2002. Seed-eaters: Seed dispersal, destruction and demography. Pp 257-271. In Levey, D.J., W.R. Silva and M. Galetti (eds.). Seed Dispersal and Frugivory: Ecology, Evolution and Conservation. CAB International.

Hunt, R., 1982. Plant growth curves. Functional approach to plant growth analysis. E. Arnold, London.

Huston, M. , 1979. A general hypothesis of species diversity. American Naturalist 113: 81-101.

IUCN, 2006. IUCN Red List of Threatened Species. <www.iucnredlist.org>.

Jansen, P.A., M. Bartholomeus, F. Bongers, J.A. Elzinga, J. ouden and S.E. Van Wieren, 2002. The role of seed size in dispersal by a scatter-hoarding rodent. Pp 209-225. In Levey, D.J., W.R. Silva and M. Galetti (eds.). Seed Dispersal and Frugivory: Ecology, Evolution and Conservation. CAB International.

Janzen, D.H., 1970. Herbivores and the number of tree species in tropical forests. American Naturalist 104: 501-528.

Janzen, D. H., 1971. Seed predation by animals. Annual Review of Ecology and Systematics 26: 465–492.

Janzen, D.H., 1986a. The future of tropical biology. Annual review of Ecology and Systematics17: 305-24.

Janzen, D.H., 1986b. Mice, big mammals, and seeds: it matters who defecates what where. Pp. 93–104 in Estrada, A. and T.H. Fleming (eds.). Frugivores and seed dispersal. Dr W. Junk Publishers, The Hague.

Janzen, D.H., 1988. management of habitat fragments in a tropical dry forest: growth. Annals of the Missouri Botanical Garden 75: 105-116.

Jensen, T.S. and O.F. Nielsen, 1986. Rodents as seed dispersers in a heath-oak wood succession. Oecologia 70: 214–221.

Jensen, T.S., 1982. Seed production and outbreaks of non-cyclic rodent populations in deciduous forests. Oecologia 54: 184–192.

Jensen, T.S., 1985. Seed-seed predator interactions of European beech, *Fagus sylvatica* and forest rodents, *Clethrionomys glareolus* and *Apodemus flavicollis*. Oikos 44: 149–156.

Jiménez, J., E. Jurado, O. Aguirre and E. Estrada, 2005. Effect of Grazing on Restoration of Endemic Dwarf Pine (*Pinus culminicola* Andresen et Beaman) Populations in Northeastern Mexico. Restoration Ecology 13: 103-107.

Jose, S., R.A. Gillespie, S.J. George and M. Kumar, 1996. Vegetation responses along edge-to-interior gradients in a high altitude tropical forest in peninsular India. Forest Ecology and Management 87: 51-62.

Kapos, V., 1989. Effects of isolation on the water status of forest patches in the Brazilian Amazon. Tropical ecology5: 173-185.

Kapos, V., E.V. Wandelli, J.L.C. Camargo, and G.M.S. Ganade, 1997. Edge-related changes in environment and plant responses due to forest fragmentation in central Amazonia. pp. 33-44. In Laurance, W.F. and R.O. Bierregaard (eds.). Tropical Forest Remnants: Ecology, Management, and Conservation of Fragmented Communities. University of Chicago Press, Chicago, Ill., USA.

Kebrom Tekle and Tesfaye Bekele, 2000. The role of soil seed bank in the rehabilitation of degraded hillslopes in Southern Wello, Ethiopia. Biotropica 32: 23–32.

Kelt, D.A., P.L. Meserve, M.L. Forister, L.K. Nabors and J.R. Gutiérrez, 2004. Seed predation by birds and small mammals in semiarid Chile. Oikos104: 133-141

Kerbs, C.J., 1999. Ecological Methodology. Second edition. Addison Wesley Longman, Inc.

Khurana, E. and J.S. Singh, 2000. Ecology of seed and seedling growth for conservation and restoration of tropical dry forest: a review. Environ. Conserv. 28: 39-52.

Kikvidze, Z., L. Khetsuriani, D. Kikodze and R.M. Callaway, 2006. Seasonal shifts in competition and facilitation in subalpine plant communities of the central Caucasus. Vegetation Science 17: 77-82.

Kitajima, K. and M. Fenner, 2000. Ecology of Seedling Regeneration. Pages 331-349. In Fenner M. (eds.), Seeds: The ecology of regeneration in plant communities. CABI publishing, 2nd Edition.

Kleinbaum, D.G. and M. Klein, 2005. Statistics for Biology and Health. Survival Analysis. A self –learning Text. Second Edition. Springer Science + Business Media, Inc. New York.

Kollmann, J. and M. Buschor, 2002. Edges effects on seed predation by rodents in deciduous forests of northern Switzerland. Plant ecology 164: 249-261.

Krebs, C.J., 1989. Ecological Methodology. Harper and Row, New York.

Krebs, C.J., 2001. Ecology: The experimental Analysis of Distribution and Abundances, 5th edn. Benjamin Cummings, San Franscisco.

Kwak, M.M., O. Velterop, and J. van Andel, 1998. Pollen and gene flow in fragmented habitats. Applied Vegetation Science 1: 37-54.

Lacoul, P. and B. Freedman, 2006. Relationships between aquatic plants and environmental factors along a steep Himalayan altitudinal gradient. Aquatic Botany 84: 3–16.

Lal, R., 1987. Tropical Ecology and Physical Edaphology. Wiley, Chichester.

Lamb, D., J. Parrotta, R. Keenan and N. Tucker, 1997. Rejoining Habitat Remnants: Restoring Degraded Rainforest Lands. Pp 366-385. In Laurance,

W. F. and R.O. Jr. Bierregaard (eds.), Tropical Forest Remnants: Ecology, Management, and Conservation of Fragmented Communities. The University of Chicago Press, Chicago & London.

Land A. D. and L. K. Rieske, 2006. Interactions among prescribed fire, herbivore pressure and shortleaf pine (*Pinus echinata*) regeneration following southern pine beetle (*Dendroctonus frontalis*) mortality. Forest Ecology and Management 235: 260-269

Landis, T.D., R.W. Tinus, S.E. McDonald and J.P. Barnett, 1989. The container Tree Nursery Manual. Agricultural Handbook. Report 674. USDA Forest Service, Washington, DC.

Laurance, W. F. and R.O. Jr. Bierregaard (eds.), 1997. Tropical forest remnants. Ecology, management, and conservation of fragmented communities. The University of Chicago Press, Chicago Chicago & London .

Laurance, W.F., 1994. Rainforest fragmentation and the structure of small mammal communities in tropical Queensland. Biological Conservation 69: 23–32.

Laurance, W.F. and E. Yensen, 1991. Predicting the impacts of edge effects in fragmented habitats. Biological Conservation 55: 77–92.

Laurent, N., and S.A.O. Chamshama, 1987. Studies on the germination of *Erythrina abyssinica* and *Juniperus procera*. Int. Tree Crops J. 4: 291–298.

Leck, M.A., V.T. Parker and R.L. Simpson (eds.), 1989. Ecology of Soil Seed Banks. Acad. Pr., San Diego.

Ledig, F.T., 1974. Concepts of growth analysis. Pp 166-182. In Reid, C.P.P. and G.H. Fechner (eds.), Proceedings of the Third North American Forest Biology Workshop, College of Forestry and Natural Resources, Colorado State University, Fort Collins, CO,

Legesse Negash, 1995. Indigenous trees of Ethiopia. Biology, uses and propagation techniques. SLU Repro. Umea, Sweden.

Legesse Negash, 2004. Rapid seed-based propagation method for the threatened African cherry (*Prunus africana*). New Forests 27: 215–227.

Lepsch-cunha, N., C. Gascon and P. Kageyama, 2001. The Genetic of Rare Tropical Trees: Implications for conservation of a demographically heterogeneous group. Pp 79-95. In Bierregaard, Jr R.O., C. Gascon, T.E. Lovejoy and R.C.G. Mesquita (eds.), Lessons from Amazonia: The ecology and conservation of fragmented forest. Yale University press, New Haven & London.

Lieberman, D., M. Li, 1992. Seedling recruitment patterns in a tropical forest in Ghana. Vegetation Science 3: 375-382.

Lieberman, D., M. Lieberman, R. Peralta, and G.S. Hartshorn, 1996. Tropical Forest Structure and Composition on a Large-Scale Altitudinal Gradient in Costa Rica. Ecology 84: 137-152.

Linhart, Y. B. and R. J. Whelan, 1980. Woodland Regeneration in Relation to Grazing and Fencing in Coed Gorswen, North Wales. Applied Ecology17: 827-840.

López-Barrera, F. and A. Newton, 2005. Edge type effect on germination of oak tree species in the Highlands of Chiapas, Mexico. Forest Ecology and Management 217: 67-79.

Logan, W.E.M., 1946. An introduction to the forests of central and southern Ethiopia. Imperial Forestry institute of Oxford Paper 24: 58 pp.

Louda, S. M., 1989. Predation in the dynamics of seed regeneration. Pp. 25–51 In: Leck M. A., V. T. Parker, R. L. Simpson (eds.), Ecology of soil seed banks. Academic Press, San Diego.

Lovett, J. C., 1996. Elevational and latitudinal changes in tree associations and diversity in the Eastern Arc mountains of Tanzania. Tropical Ecology 12: 629-650.

Lovett, J. C., 1999. Tanzanian Forest Tree Plot Diversity and Elevation. Tropical Ecology 15: 689-694.

Lugo, A.E., 1997. The apparent paradox of reestablishing species richness on degraded lands with tree monocultures. Forest Ecology and Management 99: 9-19.

MacMahon, J.A., 1997. Ecological Restoration. Pp.479-511 In Meffe, G.K. and C.R. Carroll (eds.). principles of Conservation Biology. Second edition. Sinauer Associates, Inc. Publishers, Sunderland, Massachusetts, USA.

Madsen, P., 1995. Effects of soil water content, fertilization, light, weed competition and seedbed type on natural regeneration of beech (*Fagus sylvatica*) Forest Ecology and Management 72: 251-264.

Magurran, A.E., 2004. Measuring Biological Diversity. Blackwell Publishing.

Malcolm, J.R., 1994. Edge effects in central Amazonian forest fragments. Ecology 75: 2438–2445.

Martínez-Gallardo, R. and V. Sánchez-Cordero, 1993. Dietary value of fruits and seeds to spiny pocket mice (Heteromyidae). Journal of Mammalogy 74: 436–442.

Maschinski, J., 2006. implications of population Dynamic and metapopulation theory for restoration. Pp59-87. In Falk, D.A., M.A. Palmer and J.B. Zedler (eds.), Foundations of Restoration Ecology. Society for Ecological Restoration International. Island Press, Washington.

Masresha Fetene and Yonas Feleke, 2001. Growth and photosynthesis of seedlings of four tree species from a dry tropical afromontane forest. Tropical Ecology 17: 269-283.

Matus, G., M. Papp and B. Tothmeresz, 2005. Impact of management on vegetation dynamics and seed bank formation of inland dune grassland in Hungary. Flora 200: 296-306.

Mayle, B., 1999. Domestic stock grazing to enhance woodland biodiversity, Rep. No. Information Note 28. Forestry Commission, Edinburgh.

McEvoy, P.M., J.H. McAdam, M.R. Mosquera-Losada and A. Rigueiro-Rodrí-Guez, 2006. Tree regeneration and sapling damage of pedunculate oak Quercus robur in a grazed forest in Galicia, NW Spain: a comparison of continuous and rotational grazing systems. Agroforestry System 66: 85-92.

McLaren, K.P. and M.A McDonald, 2003. The effects of moisture and shade on seed germination and seedling survival in a tropical dry forest in Jamaica. Forest Ecology and Management 183: 61-75.

McNaughton, S.J., 1983. Compensatory plant growth as a response to hervibory. OIKOS 40: 329–336.

McNaughton, S.J., 1985. Ecology of a grazing ecosystem: the Serengueti. Ecological Monographs 55: 259–294.

McRill, M. and G.R. Sagar, 1973. Earthworms and seeds. Nature (London) 243: 482.

Melaku Bekele, 1998. Forest History of Ethiopia from Early times to 1974. Walia: Journal of the Ethiopian wildlife and Natural History Society19: 3-9.

Milchunas, D.G. and I. Noy-Meir, 2002. Grazing refuges, external avoidance of herbivory and plant diversity. Oikos 99: 113-130.

Miller, A.M., C. McArthur and P.J. Smethurst, 2006. Characteristics of tree seedlings and neighbouring vegetation have an additive influence on browsing by generalist herbivores. Forest Ecology and Management 228: 197-205.

Miller, K., M.H. Allegreti, N. Johnson and B. Jonsson, 1995. Measures for Conservation of Biodiversity and Sustainable Use of its Components. Pp 915-1061. In Heywood, V.H. and R.T. Watson (eds.). Global Biodiversity Assessment. The United Nations Environment Program and Cambridge University Press, Cambridge, UK.

Molofsky, J., C.K. Augspurger, 1992. The effect of leaf litter on early seedling establishment in a tropical forest. Ecology 73: 68-77.

Montgomery, W.I. and J. Gurnell, 1985. The behavior of Apodemus. Symp. Zoological Soc. London 55: 89–115.

Morse, L.J., T.A. Day, S.H. Faeth, 2002. Effect of Netyphodium endophyte infection on growth and leaf gas exchange of Arizona fescue under contrasting water availability regimes. Environ. Exp. Bot. 48: 257-268.

Motta, R., 2003. Ungulate impact on rowan (*Sorbus aucuparia* L.) and Norway spruce (*Picea abies* (L.) Karst.) height structure in mountain forest in the eastern Italian Alps. Forest Ecology and Management 181: 139-150.

Mulugeta Lemenih and Demel Teketay, 2005. Effect of prior land use on the recolonization of
native woody species under plantation forests in the highlands of Ethiopia. Forest Ecology and Management 218: 60–73.

Mwendera , E.J. and M.A. Mohamed Saleem, 1997. Hydrologic response to cattle grazing in the Ethiopian highlands. Agriculture, Ecosystems and Environment 64 : 33-41.

Myster, R.W. and S.T.A. Pickett, 1993. Effect of litter, distance, density and vegetation patch type on post-dispersal tree seed predation in old fields. Oikos 66: 381–388.

Nason, J.D., P.R. Aldrich and J.L. Hamrick, 1997. Dispersal and the Dynamics of Genetic Structure in Fragmented Tropical Tree Populations. Pp 304-320. In Laurance, W. F. and R.O.Jr. Bierregaard (eds.), Tropical Forest Remnants: Ecology, Management, and Conservation of Fragmented Communities. The University of Chicago Press, Chicago & London.

Naveh, Z. and R.H. Whittaker. 1980. Structural and floristic diversity of shrublands and woodlands in northern Israel and other Mediterranean areas. Plant ecology 41:171–190.

Negash Mamo, Mebrate Mihretu, Miftah Fekadu, Mulualem Tigabu, and Demel Teketay, 2006. Variation in seed and germination characteristics among Juniperus procera populations in Ethiopia. Forest Ecology and Management 225: 320-327.

Nei, M., T. Maruyamo and R. Chakraborty, 1975. The bottleneck effect and genetic variation in populations. Evolution 29: 1-10.

Neptalí, R., G. Mario and W. Guadalupe, 2001. Anthropogenic disturbance and tree diversity in Montane Rain Forest in Chiapas, Mexico. Forest Ecology and Management 154: 311-326.

Nishiyama, Y., M. Moriyasu, M. Ichimaru, Y. Tachibana, A. Kato, S. G. Mathenge, J. N. Nganga and F. D. Juma, 1996. Acyclic triterpenoids from *Ekebergia capensis*. Phytochemistry 42: 803-807.

Norton, D.A., R.J. Hobbs and L. Atkins, 1995 Fragmentation, disturbance, and plant distribution – mistletoes in woodland remnants in the Western Australian wheatbelt. Conservation Biology 9: 426–438.

Notman E., D.L. Gorchov and F. Cornejo, 1996. Effect of distance aggregation and habitat on levels of seed predation for two mammal-dispersed neotropical rain forest tree species. Oecologia 106: 221–227.

Onaindia, M., I. Domingueza, I. Albizu, C. Garbisu and I. Amezaga, 2004.Vegetation diversity and vertical structure as indicators of forest disturbance. Forest Ecology and Management 195 : 341–354

Opperman, J.J. and A.M. Merenlender, 2000. Deer Herbivory as an Ecological Constraint to Restoration of Degraded Riparian Corridors. Restoration Ecology 8: 41-47.

Pammenter, N.W. and P. Berjak, 2000. Some thoughts on the evolution and ecology of recalcitrant seeds. Plant Species Biology 15: 153-156.

Pascal, J. and R. Pelissier, 1996. Structure and Floristic Composition of a Tropical Evergreen Forest in South-West India. Tropical Ecology 12: 191-214.

Passos, L. and P.S. Oliveira, 2003. Interactions between ants, fruits and seeds in a restinga forest in south-eastern Brazil. Tropical Ecology 19: 261-270.

Pedraza, R. A., and G. Williams-Linera, 2003. Evaluation of native tree species for the rehabilitation of deforested areas in a Mexican cloud forest. New Forests 26: 83-99.

Peters, C.M., 1996. The Ecology and Management of Non-Timber Forest Resources. World BankTechnical Paper Number 322.

Peterson, C.J. and J.M. Facelli, 1992. Contrasting germination and seedling growth of Betula alleghaniensis and Rhus typhina subjected to various amounts and types of plant litter. American Journal of Botany 79: 1209-1216.

Petraitis, P.S., R.E. Latham and R.A. Niesenbaum, 1989. The maintenance of species diversity by disturbance. Quarterly Review of Biology 64: 393–418.

Pfister, C.A. and M.E. Hay, 1988. Associational plant refuges: convergent patterns in marine and terrestrial communities result from differing mechanisms. Oecologia 77: 118-129.

Pohjonen Y. and T. Pukkala, 1990. *Eucalyptus globulus* in Ethiopian forestry. Forest Ecology and Management 36: 19-31.

Pohjonen Y. and T. Pukkala, 1992. *Juniperus procera* Hocht. ex. Endl. in Ethiopian Forestry. Forest Ecology and Management 49: 75-85 75 .

Poorter, L. and K. Kitajima, 2007. Carbohydrate storage and light requirements of tropical moist and dry forest tree species. Ecology 88: 1000–1011.

Powell, A.H., and G.V.N. Powell, 1987. Population dynamics of male euglossine bees in Amazonian forest fragments. Biotropica 19: 176-79.

Price, M. and S. H. Jenkins, 1986. Rodents as seed consumers and dispersers. Pp. 191-235. In Murray, D. (eds.). Seed dispersal. Academic Press, Sydney, Australia.

Pueyo, Y., C.L. Alados and C. Ferrer-Benimeli, 2006. Is the analysis of plant community structure better than common species-diversity indices for assessing the effects of livestock grazing on a Mediterranean arid ecosystem? Journal of Arid Environments 64: 698–712.

Pugnaire, F.I. and M.T. Luque, 2001. Changes in interactions along a gradient of environmental stress. Oikos 93: 42-49.

Putz, F.E., 1983. Treefall pits and mounds, buried seeds, and the importance of soil disturbance to pioneer trees on Barro Colorado Island, Panama. Ecology 64: 1069–1074.

Ramirez-Marcial, N., M. Gonzalez-Espinosa and G. Williams-Linera, 2001. Anthropogenic disturbance and tree diversity in montane rain forests in Chiapas, Mexico. Forest ecology and management 154: 311-326.

Ray, G.J. and B.J. Brown, 1995. Restoring Caribbean dry forests: evaluation of tree propagation techniques. Restoration Ecology 3: 86-94.

Rincon, E., P. Huante, 1993. Growth response of tropical deciduous tree seedlings to contrasting light conditions. Trees Struct. Funct.7: 202-207.

Ross, K.A. , B.J. Fox and D. Marilyn, 2002. Changes to plant species richness in forest fragments: fragment age, disturbance and fire history may be as important as area. Journal of Biogeography 29: 749–765

Rylands, A.B., and A. Keuroghlian, 1988. Primate population in continuous forest fragments in central Amazonia. Acta Amazónica 18: 291-320.

Sacande, M., H.W. Pritchard and A.E. Dudley, 2004. Germination and storage characteristics of Prunus africana seeds. New Forest 27: 239-250.

Sánchez-Cordero, V. and R. Martínez-Gallardo, 1998. Post-dispersal fruit and seed removal by forest-dwelling rodents in a lowland rainforest in Mexico. Tropical Ecology14: 139-151.

Sanders, H., 1968. Marine benthic diversity: a comparative study. The American Naturalist 102: 243-282.

Santos K., L.S. Kinoshita, F.A.M. dos Santos, 2007. Tree species composition and similarity in semi deciduous forest fragments of southeastern Brazil. Biological Conservation 135: 268-277

Saunders, D.A., R.J. Hobbs and C.R. Margules, 1991. Biological consequences of ecosystem fragmentation: a review. Conservation Biology 5: 12–32.

Saw, L.G., 2004. Tropical Moist Forests Encyclopedia of Forest Sciences, Pages 1775-1782. Elsevier Ltd.

Scowcroft, P. G., 1983. Tree cover changes in mamane (*Sophora chrysophylla*) forests grazed by sheep and cattle. Pac. Sci. 37: 109-119.

SGAZDA (South Gondar Zonal Department of Agriculture). General Agricultural Information Document For 2004

SGAZDPED (South Gondar Zonal Department of Plan and Economic Development). General Demographic statistic for South Gondar for 2004.
Shafer, M.L. 1981. Minimum population sizes for species conservation. Bioscience 31: 131–134.
Silvertown, J.W., 1982. Introduction to Plant Population Ecology. Longman, London–New York.
Simberloff, D., 1976. Experimental zoogeography of islands: Effects of island size.Ecology 57:629-648.
Simpson, R.L., M.A. Leck and V.T. Parker, 1989. Seed banks: general concepts and methodological issues. . Pp 3–8. In Leck, M.A., V.T. Parker and R.L. Simpson (eds.), Ecology of Soil Seed Banks. Academic Press, San Diego
Smiet, A.C., 1992. forest ecology on Java: human impact and vegetation on montane forest. Tropical Ecology 8: 129-152.
Smit, C., J.D. Ouden, H. Mueller-Schaerer, 2006. Unpalatable plants facilitate tree sapling survival in wooded pastures. Applied Ecology 43: 305-312.
Smit, C., M. Gusberti and H. Mueller-Schaerer, 2006. Safe for saplings; safe for seeds? Forest Ecology and Management.,doi:10.1016/j.foreco.2006.09.069.
Smith, C. C. and O. J. Reichman, 1984. The evolution of food caching by birds and mammals. Annu. Rev. Ecol. Syst. 15: 329–351.
Smythe, N., 1986. Competition and resource partitioning in the guild of neotropical terrestrial frugivorous mammals. Annual Review of Ecology and Systematics 17: 168–188.
Society for Ecological Restoration International Science & Policy Working Group. 2002. The SER Primer on Ecological Restoration. www.ser.org.
Soulé , M.E., A.C. Alberts, and D.T. Bolger, 1992. The effects of habitat fragmentation on chaparral plants and vertebrates. Oikos 63: 39–47.
Stevens, G.C., 1992. The Elevational Gradient in Altitudinal Range: An Extension of Rapoport's Latitudinal Rule to Altitude. The American Naturalist 140: 893-911.
Stone, C.P., 1985. Alien animals in Hawaii's native ecosystems: toward controlling the adverse effects of introduced vertebrates. Pp. 252–297. In Stone, C.P., and J. M. Scott (eds.). Hawaii's terrestrial ecosystems: preservation and management,. Cooperative National Park Resources Study Unit, University of Hawaii, Honolulu, Hawaii.
Sullivan, T.P., 1978. Biological control of conifer seed damage by the deer mouse (*Peromyscus maniculatus*). Proceedings 8th Vertebrate Pest Conference, Sacramento, California, pp. 237-250.

Sullivan, T. P., 1979. The Use of Alternative Foods to Reduce Conifer Seed Predation by the Deer Mouse (*Peromyscus maniculatus*). Applied Ecology 16: 475-495.

Sun, D., G.R. Dickinson and A.L. Bragg, 1995. Direct seeding of *Alphitonia petriei* (Rhamnaceae) for gully revegetation in tropical northern Australia. Forest Ecology and Management 73: 249-257.

Sun, D., G.R. Dickinson and A.L. Bragg, 1997. Effect of cattle grazing and seedling size on the establishment of *Araucaria cunninghamii* in a silvo-pastoral system in northeast Australia. J. Environ. Manag. 49: 435-444.

Tadesse Woldemariam, 1998. Diversity of Woody Plants and Avifauna in a Dry Afromontane Forest, Onthe Central Plateau of Ethiopia. SLU: Ethiopian MSc in Forestry Program thesis work.

Tamirat Bekele, 1994. Studies on remnant Afromontane forests on the Central Plateau of Shewa, Ethiopia. Ph.D. thesis. Acta Unifersitatis Upsaliensis, Uppsala.

Taylor, D. A. H., 1981. Ekebergin, a limonoid extractive from *Ekebergia capensis*. Phytochemistry 20: 2263-2265.

Teague, W.R., 1989. The rate of consumption of bush and grass by goats in a representative Acacia karroo savanna community in the Eastern Cape. Journal of Grassland Soc. South. Afri 6: 8-13.

Tefera Mengistu, 2001. The role of area closures in the recovery of woody vegetation in degraded hillsides of Biyo and Tiya, central and northern Ethiopia. Swedish university of agricultural sciences.

Tefera Mengistu, Demel Teketay, H. Hakan and Yonas Yemshaw, 2005. The role of enclosures in the recovery of woody vegetation in degraded dryland hillsides of central and northern Ethiopia. J. Arid Environ. 60: 259–281.

Teich, I., A.M. Cingolani, D. Renison, I. Hensen, M.A. Giorgis, 2005. Do domestic herbivores retard Polylepis australis Bitt. woodland recovery in the mountains of Córdoba, Argentina? Forest Ecology and Management 219: 229-241.

Tesfaye Wubet, Ingrid Kottke, Demel Teketay, Franz Oberwinkler, 2003. Mycorrhizal status of indigenous trees in dry Afromontane forests of Ethiopia. Forest ecology and Management 179: 387-399.

Thompson, K. and J.P. Grime, 1979. Seasonal variation in the seed bank of herbaceous species in ten contrasting habitats. Ecology 67: 893–921.

Thompson, K., 1987. Seeds and seed banks. New Phytol. 106 (Suppl.): 23–34.

Thompson, K., 1992. Functional Ecology of soil seed banks. Pp. 231-258. In Fenner, M. (ed.). Seeds: the ecology of regeneration in plant communities. CAB International, Wallingford.

Thompson, K., J.p. Grime, and G. Mason, 1977. Seed germination in response to diurnal fluctuations of temperature. Nature 267: 147-149.

Tuomisto, H., K. Ruokolainen and M. Yli-Halla, 2003. Dispersal, Environment, and Floristic variation of Western Amazonian Forests. Science 299: 241-244.

Turner, I.M. 1996. Species loss in fragments of tropical rain forest – a review of the evidence. Applied Ecology 33: 200–209.

Turton , S.M. and H.J. Freiburger, 1997. Edge and aspect effects on the microclimate of a small tropical forest remnant on the Atherton Tableland, northeastern Australia. Pp.45-54. In Laurance, W.F. and R.O.Jr. Bierregaard (eds.), Tropical Forest Remnants: Ecology, Management, and Conservation of Fragmented Communities. University of Chicago Press, Chicago.

Vallejo, R., J. Aronson, J.G. Pausa and J. Cortina, 2006. Restoration of Mediterranean woodlands. Pp193-207. In Van Andel, J and J. Aronson (eds.), Restoration Ecology. Blackwell Publishing.

Vallentine, J.F., 2001. Grazing management. 2nd edition. Academic Press, San Diego, 659 pp.

Van Andel, J and J. Aronson, 2006. Restoration Ecology. Blackwell Publishing.

Van Andel, J. and A.P. Grootjans, 2006. Concepts in restoration ecology. Pp16-28. In Van Andel, J and J. Aronson (eds.), Restoration Ecology. Blackwell Publishing.

Van Andel, J., 2006. Populations: intraspecific interactions.Pp70-81. In Van Andel, J and J. Aronson (eds.), Restoration Ecology. Blackwell Publishing.

Van Bloem, S.J., P.G. Murphy and A.E. Lugo, 2004. Tropical Dry Forests. Encyclopedia of Forest Sciences, Pages 1767-1775. Elsevier Ltd.

Van den Berg, L.J.L., J.M. Bullock, R.T. Clarke, R.H.W. Langston and R.J. Rose, 2001. Territory selection by the Dartford warbler (Sylvia undata) in Dorset, England: the role of vegetation type, habitat fragmentation and population size. Biological Conservation 101: 217–28.

Van der Valk, A.G. and R.L. Pederson, 1989. Seed banks and the management and restoration of natural vegetation. In: Leck, M.A., V.T. Parker, R.L. Simpson (eds.), Ecology of Soil Seed Banks. Academic Press, San Diego, pp. 329–346.

Vazquez, G., J.A. and T. J. Givnish, 1998. Altitudinal Gradients in Tropical Forest Composition, Structure, and Diversity in the Sierra de Manantlan. Ecology 86: 999-1020.

Vazquez-Yanes, C. and A. Orozco-Segovia, 1993. Patterns of seed longevity and germination in the tropical rain forest. Annual Review of Ecology and Systematics 24: 69-87.

Vazquez-Yanes, C. and A. Orozco-Segovia, 1994. Signals for seeds to sense and respond to gaps. Pp.209-236. In Caldwell, M.M. and R.W. Pearcy (eds.),

Exploitation of environmental heterogeneity by plants: ecophysiological processes above and below ground. Academic Press, Inc., New York.

Viana, V.M., A. A.J. Tabanez and J.L.F Batista, 1997. Dynamics and Restoration of Forest Fragments in The Brazilian Atlantic Moist Forest. Pp 351-365. In Laurance, W. F. and R.O. Jr. Bierregaard (eds.), Tropical Forest Remnants: Ecology, Management, and Conservation of Fragmented Communities. The University of Chicago Press, Chicago & London.

Vitousek, P.M. and J.S. Denslow, 1986. Nitrogen and Phosphorous availability in tree fall gaps of a lowland tropical rainforest. Ecology 74: 1167-1178.

Walker, L.R. and R.D. Moral, 2003. Primary Succession and Ecosystem Rehabilitation. Cambridge university press.

White, F., 1983. The vegetation of Africa. A descriptive memoir to accompany the UNESCO/AETFAT/UNSO Vegetation map of Africa. Paris, UNESCO, 356pp

Whitmore, T.C., 1991. An introduction to tropical rain forests. Oxford University Press, Oxford. 226pp.

Wijdeven, S.M.J. and M.E. Kuzee, 2000. Seed availability as a limiting factor in forest recovery processes in Costa Rica. Restoration Ecology 8: 414–424.

Wilcove, D.S., C.H. McLellan and A.P. Dobson, 1986. Habitat fragmentation in the temperate zone. Pp 237–56. In Soulé, M.E. and M.A. Sunderland (eds.), Conservation Biology. Sinauer.

Wilkinson, G.R., W.A. Nielsen, 1995. Implications of early browsing damage on the long term productivity of eucalypt forests. Forest Ecology and Management 74: 117-124.

Williams-Linera, G., 1990. Vegetation structure and environmental conditions of forest edges in Panama. Ecology 78: 356-373.

Wong, T.C.M., N.S. Sodhi and I.M. Turner, 1998. Artificial nest and seed predation experiments in tropical lowland rainforest remnants of Singapore. Biological Conservation 85: 97–104.

Workneh Gebresilassie, Afework Bekele, Gurja Belay and M. Balakrishnan, 2004. Microhabitat choice and diet of rodents in Maynugus irrigation field, northern Ethiopia. African journal of Ecology 42: 315-321.

Yates, C.J., D.A. Norton and R.J. Hobbs, 2000. Grazing effects on plant cover, soil and microclimate in fragmented woodlands in south-western Australia: implications for restoration. Austral Ecology 25: 36–47.

Young, A., T. Boyke and T. Brown, 1996. The population genetic consequences of habitat fragmentation for plants. Trends Ecolo.Evol. 11: 413-18.

Young, K.R., 1985. Deeply buried seeds in a tropical wet forest in Costa Rica. Biotropica 17: 336–338.

Zar, J.H., 1984. Biostastical analysis. Prentice-Hall, Englewood cliffs, New Jersey.

APPENDIX

LIST OF WOODY SPECIES FOUND IN THE 28 CHURCH FORESTS

Annex 1 . List of woody species found in the 28 church forests with their respective Importance value index (IVI), Relative basal area (RBA) (calculated from individuals with ≥5cm dbh), Relative density (RD) (number of individuals of a species/total individuals enumerated) and number of forests in which the species was found. IVI, RBA and RD for each species are sum of the values from all 28 forests.

Ser No	Species	Family	Life form	IVI pooled from 28 forests	RBA pooled from 28 forests	RD pooled from 28 forests	Number of forests occurred
1	*Abutilon figarianum* Webb	Malvaceae	Shrub	10	0.03	12.58	6
2	*Acacia abyssinica* Hochst. ex. Benth.	Fabaceae	Tree	20.7	29.78	13.56	15
3	*Acacia brevispica* Harms	Fabaceae	Tree	1.2	1.73	0.69	2
4	*Acacia lahai* steud. & Hochst.ex Benth.	Fabaceae	Tree	0.4	0.08	0.48	1
5	*Acacia nilotica* (L.) Willd. Ex Del.	Fabaceae	Shrub	0.5	0.00	0.60	1
6	*Acacia polyacantha* Willd.	Fabaceae	Tree	0.5	0.42	0.43	1
7	*Acanthus pubescens* (Oliv.) Engl.	Acanthaceae	Shrub	30.6	0.32	39.91	17
8	*Acanthus sennii* Chiov.	Acanthaceae	Shrub	2	0.00	2.61	1
9	*Acokanthera schimperi* (DC.)Benth.	Apocynaceae	Shrub	64.8	20.78	107.78	10
10	*Adansonia digitata* L.	Bombacaceae	Tree	1.3	1.51	0.99	1
11	*Albizia schimperiana* Oliv.	Fabaceae	Tree	38.9	64.41	22.22	14
12	*Albizia anthelmintica* (A.Rich.) Brogn.	Fabaceae	Tree	0.5	0.00	0.59	1
13	*Allophylus abyssinica* (Hochst) Radlkofer	Sapindaceae	Tree	0.7	0.41	0.51	2
14	*Apodytes dimidiata* E. Mey ex. Arn.	Icacinaceae	Tree	20.9	33.15	13.19	7
15	*Arundinaria alpina* K.Schum.	Gramineae	Stemmed grass	1.1	0.15	0.95	3
16	*Arundo donax* L.	Gramineae	Stemmed grass	4.7	0.00	7.84	7
17	*Asparagus aethiopicus* L.	Liliaceae	Shrub	4.4	0.00	4.56	7
18	*Asplenium trichomanes* L.	Aspliniaceae	Shrub	0	0.04	3.40	2
19	*Bersama abyssinica* Fresen.	Melianthaceae	Tree	36.8	3.45	55.88	16
20	*Biophytum abyssinicum* Steud. Ex Rich	Oxalidaceae	Shrub	0.3	0.10	0.30	1
21	*Bridelia micrantha* (Hochst.) Baill.	Euphorbiaceae	Shrub	0	3.59	5.87	7
22	*Brucea antidysenterica* J.F.Mill.	Simaroubaceae	Shrub	26.8	0.44	28.60	17
23	*Buddleja polystachya* Fresen.	Loganiaceae	Tree	30.5	43.79	19.46	16
24	*Cajanus cajan* (L.) Millsp.	Fabaceae	Shrub	0.5	0.00	0.74	1
25	*Calpurnia aurea* (Ait.) Benth.	Fabaceae	Shrub	65.9	3.25	98.69	22

26	*Canthium oliocarpum* Hiern	Rubiaceae	Shrub	20.3	2.53	24.60	13
27	*Capparis tomentosa* L.	Capparidaceae	Climber	29.3	1.44	34.78	17
28	*Carissa edulis* Vahl.	Apocynaceae	Shrub	86	6.11	124.88	25
29	*Casuarina equisetifolia* L.	Casuarinaceae	Tree	0.6	0.00	1.10	1
30	*Celtis africana* Burm.f.	Ulmaceae	Tree	32.3	45.50	22.47	11
31	*Chionanthus mildbraedii* (Gilg & Schellenb.) Stearn	Oleaceae	Tree	20.4	35.89	10.76	2
32	*Cissus quadrangularis* L.	Vitaceae	Climber	4.4	0.00	5.04	4
33	*Citrus aurantifolia* (Christm.)	Rutaceae	Tree	1.1	0.09	1.29	2
34	*Citrus aurantium* L.	Rutaceae	Tree	0.3	0.12	0.30	1
35	*Clausena anisata* (Willd.)Benth.	Rutaceae	Tree	29.3	4.59	35.22	19
36	*Clematis hirsuta* Perr & Guill.	Ranunculaceae	Climber	5.1	0.08	5.24	9
37	*Clerodendron myricoides* (Hoechst)R.Br.ex.Vatke	Verbenaceae	Shrub	11.1	0.37	14.00	8
38	*Clutia abyssinica* Jaub. & Spach.	Euphorbiaceae	Shrub	18.5	0.00	22.26	14
39	*Coffea arabica* L.	Rubiaceae	Shrub	13.8	1.33	27.56	3
40	*Colutea abyssinica* kunth & Bouche	Fabaceae	Shrub	1	0.00	1.42	1
41	*Combretum adenogonium* Steud. Ex A. Rich	Combretaceae	Tree	1.1	1.31	0.82	2
42	*Combretum collinum* Fresen.	Combretaceae	Tree	0.9	0.79	0.78	1
43	*Combretum molle* R.Br.ex G.Don	Combretaceae	Tree	16.1	7.60	22.43	8
44	*Cordia africana* Lam.	Boraginaceae	Tree	15.5	24.24	9.33	8
45	*Croton macrostachyus* Del.	Euphorbiaceae	Tree	24.7	21.97	22.25	18
46	*Cupressus lusitanica* Mill.	Cupressaceae	Tree	6.4	2.24	12.41	5
47	*Cyphostemma adenocaule* (Steud. Ex A. Rich.) Descoings	Vitaceae	Climber	6	0.73	6.63	3
48	*Cyphostemma molle* (Bak.) Descoings	Vitaceae	Climber	0.2	0.00	0.24	1
49	*Dalbergia lactea* Vatke	Fabaceae	Shrub	0.3	0.01	0.30	1
50	*Delphinium wellbyi* Hemsl	Ranunculaceae	Tree	0.8	0.54	0.78	1
51	*Dichrostachys cinerea* (L.) Wight &Arn.	Mimosoideae	Tree	0.5	0.00	0.63	2
52	*Discopodium penninervium* Hochst.	Solanaceae	Shrub	5.8	0.65	5.97	7
53	*Diospyros abyssinica* (Hiern)F.White	Ebenaceae	Tree	21	4.75	36.24	3
54	*Diospyros mespiliformis* A. Dc.	Ebenaceae	Tree	0.7	0.61	0.63	2
55	*Dodonaea anguistifolia* L.f.	Sapindaceae	Shrub	43.1	0.87	73.39	12
56	*Dolichos sericeus* E. Mey.	Fabaceae	Climber	0.4	0.15	0.30	1
57	*Dombeya torrida* (J.F. Gmel.) P.Bamps	Sterculiaceae	Tree	17.8	7.79	20.25	14
58	*Dovyalis abyssinica* (A.Rich.)Warb.	Flacourtiaceae	Tree	24	18.81	22.64	10
59	*Dracaena steudneri* Schweinf.ex Engl.	Agavaceae	Tree	23.7	25.06	21.42	11
60	*Ekebergia capensis* Sparrm.	Meliaceae	Tree	38.7	89.00	10.07	12
61	*Embelia schimperi* Vatke	Myrsinaceae	Climber	4.1	0.13	4.34	5
62	*Erica arborea* L.	Ericaceae	Tree	34.2	10.06	58.80	4

63	*Eruthrina brucei* Schweinf.	Fabaceae	Tree	22.9	18.68	25.09	9
64	*Eucalyptus camaldulensis* Dehnh.	Myrtaceae	Tree	22.4	7.46	45.05	9
65	*Eucalyptus citriodora* Hook	Myrtaceae	Tree	0.6	0.00	0.79	1
66	*Eucalyptus globulus* Labill.	Myrtaceae	Tree	19.4	22.62	27.29	6
67	*Euclea divinorum* Heirn	Ebenaceae	Shrub	27.2	0.70	37.29	10
68	*Euphorbia abyssinica* Gmel.	Euphorbiaceae	Tree	112.6	152.68	97.79	19
69	*Euphorbia polyacantha* Boiss	Euphorbiaceae	Shrub	1.3	0.00	1.14	1
70	*Euphorbia tirucalli* L.	Euphorbiaceae	Shrub	0.9	0.00	0.94	1
71	*Ficus ingens* (Miq.) Miq.	Moraceae	Tree	2.3	4.47	1.09	2
72	*Ficus ovata* (Vahi)	Moraceae	Tree	0	8.58	4.33	8
73	*Ficus sur* Forssk.	Moraceae	Tree	2	2.94	1.15	4
74	*Ficus sycomorus* L.	Moraceae	Tree	2.6	4.23	1.51	3
75	*Ficus thonningii* Blume	Moraceae	Tree	43.5	102.34	10.15	9
76	*Ficus vallis-choudae* Del.	Moraceae	Tree	10.5	22.17	4.53	2
77	*Ficus vasta* Forssk.	Moraceae	Tree	41.4	109.79	4.39	9
78	*Ficus carica* L.	Moraceae	Tree	1.2	1.20	0.87	2
79	*Ficus ingens* (Miq.) Miq.	Moraceae	Tree	0.2	0.03	0.24	1
80	*Ficus sycomorus* L.	Moraceae	Tree	0.3	0.21	0.24	1
81	*Galiniera saxifraga* (Hochst.) Bridson	Rubiaceae	Shrub	1.4	0.58	1.30	3
82	*Gardenia ternifolia* Schumach. & Thonn.	Rubiaceae	Tree	0.4	0.05	0.49	2
83	*Gladiolus psittacinus* Hook.	Simaroubaceae	Climber	0.3	0.00	0.39	1
84	*Gnidia glauca* (Fresen.) Gilg	Thymelaeaceae	Tree	2.4	1.73	1.77	3
85	*Grewia ferruginea* Hochst.ex.A.Rich	Tiliaceae	Shrub/Small tree	16.8	3.05	18.87	13
86	*Grewia bicolor* Juss	Tiliaceae	Tree	0.2	0.06	0.24	1
87	*Hagenia abyssinica* (Bruce) J.F.Gmelin	Rosaceae	Tree	2.6	1.49	1.48	2
88	*Hippocratea africana* (Willd.)Loes	Celastraceae	Climber	0.8	0.00	0.99	1
89	*Hypericum revolutum* Vahl	Hypericaceae	Shrub	3.6	0.71	3.23	4
90	*Ilex mitis* (L.) Radlk.	Aquifoliaceae	Tree	2.9	0.09	4.30	1
91	*Jasminum abyssinicum* Hochst. Ex DC.	Oleaceae	Climber	1.6	0.31	1.78	3
92	*Jasminum grandiflorum* L.	Oleaceae	Shrub	15.2	1.04	17.69	13
93	*Juniperus procera* L.	Cupressaceae	Tree	460.4	849.97	313.15	24
94	*Justicia schimperiana* (Hochst. Ex Nees) T.Anders.	Acanthaceae	Shrub	65.9	0.00	111.82	18
95	*Maesa lanceolata* Forsk	Myrsinaceae	Shrub	3.8	0.91	4.58	7
96	*Maytenus arbutifolia* (A. Rich.) Wilczek	Celastraceae	Shrub	115.1	0.92	164.03	27
97	*Maytenus senegalensis* (Lam.) Exell	Celastraceae	Tree	32.4	29.75	27.36	12
98	*Maytenus undata* (Thunb.) Blakelock,	Celastraceae	Shrub	1.2	0.00	1.56	1
99	*Melia azedarach* L.	Meliaceae	Tree	0.4	0.00	0.37	1

100	*Millitea ferruginea* (Hochst.) Bak.	Fabaceae	Tree	25.6	27.75	20.24	12
101	*Mimusops kummel* Bruce ex DC.	Sapotaceae	Tree	75	148.62	50.90	9
102	*Myrica salicifolia* A.Rich.	Myricaceae	Tree	3.1	4.81	1.37	4
103	*Myrsine africana* L.	Myrsinaceae	Shrub	18.7	0.00	28.42	9
104	*Nuxia congesta* R. Br. ex Fresen.	Loganiaceae	Tree	14.9	10.43	13.78	12
105	*Olea capensis* L.	Oleaceae	Tree	0.5	0.31	0.45	1
106	*Olea europaea* L.	Oleaceae	Tree	181.8	364.62	78.63	18
107	*Olinia rochetiana* A.Juss.	Oliniaceae	Tree	7.6	13.41	5.11	1
108	*Opuntia ficus-indica* (L.) Miller.	Cactaceae	Tree	2.4	0.67	3.83	2
109	*Osyris quadripartita* Decn.	Santalaceae	Shrub	33.9	8.52	38.31	19
110	*Otostegia tomentosa* A.Rich.	Lamiaceae	Shrub	4.5	0.00	6.01	5
111	*Oxygonum sinuatum* (Mesin.) Dammer	Polygonaceae	Shrub	0.2	0.00	0.15	1
112	*Pavetta oliveriana* Hiern	Rubiaceae	Tree	1.8	1.09	2.36	1
113	*Phoenix reclinata* Jacq.	Palmae	Tree	2.3	0.00	3.54	3
114	*Phyllanthus ovalifolius* Forssk.	Euphorbiaceae	Shrub	1.2	0.00	1.02	1
115	*Phytolacca dodecandra* L'H' erit.	Phytolaccaceae	Climber	11.2	1.39	11.98	19
116	*Piliostigma thonningii* (Schum.) Milne-Rech.	Caesalpinioideae	Tree	0.5	0.60	0.39	1
117	*Pittosporum spp.*	Pittosporaceae	Tree	2.9	2.67	1.81	4
118	*Pittosporum viridiflorum* Sims	Pittosporaceae	Tree	15.5	22.92	8.66	6
119	*Podocarpus falcatus* (Thunb.)Mirb.	Podocarpaceae	Tree	14.8	33.40	4.30	3
120	*Premna schimperi* Engl.	Verbenaceae	Shrub	16.8	4.97	20.92	9
121	*Protea gaguedi* J.F. Gmel.	Proteaceae	Tree	0.4	0.19	0.43	1
122	*Prunus africana* (Hook. f .) Kalkm.	Rosaceae	Tree	34.2	30.95	45.23	9
123	*Psydrax schimperiana* (A. Rich.) Bridson	Rubiaceae	Tree	10.5	8.78	14.11	6
124	*Pterollobium stellatum* (Forssk.) Brenan	Fabaceae	Climber	27.6	1.09	33.66	15
125	*Rhamnus prinoides* L'Herit.	Rhamnaceae	Shrub	5	0.37	8.92	6
126	*Rhamnus staddo* A. Rich.	Rhamnaceae	Tree	2.2	0.46	1.97	1
127	*Rhus retinorrhoea* Oliv.	Anacardiaceae	Tree	6.9	1.37	6.92	7
128	*Rhus glutinosa* A.Rich.	Anacardiaceae	Tree	45.7	43.91	40.98	19
129	*Rhus quartiniana* A Rich	Anacardiaceae	Tree	5.9	1.59	6.89	5
130	*Rhus vulgaris* Meikle	Anacardiaceae	Tree	9	4.23	9.34	8
131	*Ricinus Communis* L.	Euphorbiaceae	Shrub	0.6	0.00	0.60	2
132	*Ritchiea albersii* Gilg	Capparidaceae	Tree	23.1	16.27	24.40	12
133	*Rosa abyssinica* Lindley	Rosaceae	Climber	17.8	0.83	16.66	12
134	*Rothmannia urcelliformis* (Hiern) Robyns	Rubiaceae	Tree	4.7	2.04	6.21	2
135	*Rumex nervosus* Vahl.	Polygonaceae	Shrub	8.2	0.00	11.07	10
136	*Sapium ellipticum* (Hochst.)pax.	Euphorbiaceae	Tree	6.3	14.43	2.25	2

137	*Sarcostemma viminale* (L.)R.Br.	Solanaceae	Climber	4.8	0.23	5.76	3
138	*Schefflera abyssinica* (Hochst.ex.A.Rich.)Harms	Araliaceae	Tree	22.8	45.40	8.63	11
139	*Schrebera alata* (Hochst.) Baill.	Oleaceae	Tree	0	30.63	15.33	8
140	*Securinega virosa* (Roxb.ex Willd) Pax.&Hoffm	Euphorbiaceae	Tree	1.3	0.37	1.39	2
141	*Senecio gigas* Vatke	Asteraceae	Shrub	2.1	0.15	2.55	1
142	*Senna singueana* (Del.)Lock	Fabaceae	Shrub	2.2	0.02	2.52	5
143	*Senna petersiana* (Bolle) Lock	Fabaceae	Tree	1.5	0.08	1.79	1
144	*Sesbania sesban* (L.) Merrill	Papilionoideae	Shrub	0.6	0.00	1.20	1
145	*Solanecio gigas* (Vatke) C. Jeffrey	Urticaceae	Shrub	0.5	0.00	0.82	1
146	*Solanum giganteum* Jacq.	Solanaceae	Shrub	0.8	0.00	0.90	1
147	*Steganotaenia araliacea* Hochst.	Apiaceae	Tree	0.5	0.23	0.45	1
148	*Stephania abyssinica* (Dillon &A.Rich.) Walp.	Menispermaceae	Climber	12.6	1.75	11.51	12
149	*Stereospermum kunthianum* Cham.	Bignoniaceae	Tree	0.9	0.11	1.06	3
150	*Syzygium guineense* F.white	Myrtaceae	Tree	13.1	29.58	4.20	6
151	*Teclea nobilis* Del.	Rutaceae	Tree	70.1	30.98	101.87	14
152	*Urera hyselodendron* (A.Rich.) Wedd.	Urticaceae	Climber	15.4	6.10	15.02	12
153	*Vernonia amygdalina* Del.	Compositae	Tree	2.9	0.51	2.76	5
154	*Vernonia myriantha* Hook.f.	Compositae	Shrub	28	0.75	33.63	17
155	*Woodfordia uniflora* (A. Rich.) Koehne	Lythraceae	Tree	0.3	0.03	0.39	1
156	*Ximenia americana* L.	Olacaceae	Tree	3.5	1.09	3.81	4
157	*Zizyphus mucronata* Willd.	Rhamnaceae	Tree	0.2	0.03	0.24	1
158	*Unidentified* (Akenedaba)		Tree	4.2	4.77	3.21	2
159	*Unidentified* (Amago)		Tree	0.9	0.92	0.95	1
160	*Unidentified* (Aybesh)		Tree	0.3	0.29	0.24	1
161	*Unidentified* (Dihana)		Shrub	0.3	0.00	0.30	1
162	*Unidentified* (Doro Kus)		Tree	0.3	0.12	0.39	1
163	*Unidentified* (Fren)		Tree	0.4	0.08	0.48	1
164	*Unidentified* (Jigra Lab)		Shrub	0.5	0.04	0.60	1
165	*Unidentified* (Jingurit)		Tree	1.1	0.44	1.17	1
166	*Unidentified* (Mar-Hareg)		Climber	0.6	0.00	0.71	1
167	*Unidentified* (Shibre)		Tree	0.3	0.28	0.24	1
168	*Unidentified* (Worchebo)		Tree	0.9	0.78	0.78	1

SUMMARY

In Northern Ethiopia almost all dry Afromontane forests have been converted to open agricultural lands. Only small isolated fragments remain around churches ("church forests"), but these are many. This study analyses forest community structure and composition of the church forests, investigates major bottlenecks for regeneration of woody species, and explores opportunities and challenges for restoration. In this thesis, the following major questions are addressed:

1. How do forest structure, species composition and biodiversity vary across church forests and what are the major factors driving such variations? Factors considered are altitude, forest size and human influence.

2. What are the major bottlenecks in the regeneration of woody plants in church forests? These bottlenecks are studied for one to seven church forests and a major focus is given to the effects of soil seed bank and post-dispersal seed predation on seedling establishment, and of livestock grazing, microsite gradients and management interventions on seedling establishment, seedling survival and growth.

The species and structural composition of 28 forests located at different altitudes (range 1816 to 3111 masl) and of various sizes (range 1.6 to 100 ha) was assessed, in relation to altitude, forest area, livestock grazing intensity and wood harvest intensity. A total of 168 woody species (100 tree species, 51 shrubs and 17 lianas) representing 69 families were recorded in the 28 church forests studied. Out of these species, 160 were indigenous and only eight were exotic (2 shrub and 6 tree species) representing 6 families. These forests accommodate many species represented by single individuals (rare species) and also many species found only in single plots (unique), which makes church forests priority of conservation efforts. Forests differed strongly in species number (15 to 78), basal area (4.8 to 111.5 m^2/ha), density ($\geq$ 5 cm dbh: 267 to 1553/ha; >1cm diameter: 619 to 2421/ha and; seedlings: 0 to 5263/ha).

Altitude is the main factor determining species composition of these forests. Our results showed that altitudinal gradient is the main determinant of differences in species composition among these forests. Maximum similarity in species composition was found with minimum altitude difference between forests.

Geographical distance had only a weak effect on similarity. Therefore their vast altitudinal distribution gives these forests the opportunity to hold most of the biodiversity resources of the area. This can be confirmed by the number of species we found in our 28 church forests (168) which is more than the expected number of species (125 woody species) listed in the study area by South Gondar Zonal Department of Agriculture. Structural composition of these forests, on the other hand, is determined by human influence. In the understorey the interaction effect of altitude with human influence (particularly cattle interference) determines the species composition. Not only the species composition but also the number of seedlings were severely affected by cattle interference and few, if any, seedlings were found in some of the forests. In the long run this would probably lead to a human effect on overstorey composition of these forests as well. Forest area did not show a significant effect. This implies that although large sized forests are a necessary element of successful reproduction of woody species, small patches and appropriate matrix management could be useful complements for biodiversity conservation. For many tree species regeneration in most of these forests is very limited. Possible reasons for this limitation are addressed in field experiments.

Soil seed bank analysis on seven of these forests showed that these forests accumulate large quantities of persistent seeds of herbaceous species in the soil, but only five (6%) of the 91 woody species recorded in the standing vegetation of the seven forests were represented in their soil seed banks. Most of the tree species do not accumulate seeds in the soil. In order to investigate the long term behavior of seeds in forest soil, we assessed seed viability of five tree species in four sites of one forest after being buried 6, 12, or 18 months. Seed viability decreased sharply with burial time in soil for all species except for seeds of *Juniperus*, which still had 91% of viability after 18 months. Species significantly affect the viability of the seeds after 18 months of burial. The quick decline of seed viability in forest soil indicates that the study species do not have a persistent soil seed bank. The fact that most of the dominant tree species do not accumulate seeds and maintain viability in the soil suggests that their regeneration from seeds would be prevented by removal of mature individuals in the standing vegetation. The experiment of post-dispersal seed predation on seeds of six tree species in one forest showed that 92% of the seeds were predated within 3.5 months. Therefore lack of persistent soil seed bank aggravated by intense seed predation undermines seed availability for regeneration in church forests.

For four selected tree species in two forests this study showed that livestock grazing is one of the main bottlenecks hampering seedling establishment, seedling survival and seedling growth. Almost none of the sown seeds were able to germinate in unfenced plots (in unfenced plots germination was 4 and 5%, compared to 57.5 and 63.3% in fenced plots, data for Dengolt and Gelawdios respectively). In the fenced plots, seedling survival was higher (Dengolt = 65 and Gelawdios = 56%) and seedlings grew faster while there was no survival in unfenced plots. This implies that controlling livestock grazing is of paramount importance for both the internal regeneration of church forests and for restoration of the degraded surroundings.

The study also explored how tree regeneration varies along the gradient from the forest interior to the edge and open fields, and differs between canopy gaps and closed canopy sites inside the forest. Seedling establishment was more successful inside the forest and in particular in the gaps within the forest. The exposure effect between forest interior and exterior was not analogous to the exposure effect of gaps inside the forest. This result suggests that germination may be primarily influenced by moisture availability along the forest interior-open field gradient. Though seedling establishment is higher inside the forest, seedlings grew more rapidly and survived better on the outer edge of the forest. Our result revealed that the negative effect of edge on regeneration is seen on the inner edge while the outer edge has positive effect on seedlings survival. As a result of the interplay effects of light and moisture, the outer forest edge might become the optimal place for seedling survival and growth. This may indicate that seedlings can colonize and restore the forests on the degraded land from immediate surrounding areas of the church forests, given that the surrounding land is protected from grazing intervention and farming. Within the forest, light in gaps favored survival and growth of seedlings both in dry and wet periods. Perhaps, the soil moisture in gaps inside the forest was not as depleted as that outside the forest in the dry period. In general, the quantitative effect of microsite differed with species.

Restoration experiments using management interventions (sowing seeds, planting seedlings, weeding, litter removal, soil scarification) showed that the combined effect of seed sowing and litter removal increased seedling establishment significantly for all the species. This implies that insufficient availability of seeds could be one of the constraints for regeneration in church forests. Weeding did not improve seedling survival and growth of the species, and had even a negative effect on some of them.

The future existence of the woody flora and vegetation characteristic of dry Afromontane areas in Ethiopia depends on effective conservation and sustainable utilization of the remnant natural forest patches. If not possible to conserve all remnant forests, selection of church forests across altitudinal variation is important to accommodate most of the species diversity of the area. Maintaining viable populations in the forests and providing connections between forests is pivotal in this respect. Excluding cattle interference and reducing intensity of wood harvest are a prerequisite to facilitate regeneration in church forests. To offset the seed limitation sowing seeds combined with litter removal and slight soil scarification can be very useful. Sowing of seeds should be integrated with active reduction of predation and herbivory (e.g. in the form of enclosures and providing alternative food sources for predators). Seedling transplanting is another alternative to overcome some of the bottlenecks of germination especially moisture deficit and intense seed predation, but weeding might not be needed. Within the forest, opening up of the canopy may be required to facilitate seed germination and seedling performance depending on the species. Along the gradient from forest interior to edge and open field, seed sowing will be effective in the interior microsite, while transplanting seedling in the outer forest edge gives better survival and growth. These measures improve tree regeneration. Interconnecting these remnant forests by vegetation corridors following natural terrain or stream lines, or reducing the distance between them by creating buffer areas and plantations around them, and developing more patches in the landscapes are possible management activities. These will facilitate propagule and germplasm flow and ultimately may sustain these forests and help restoring the surrounding landscape.

SAMENVATTING

Nagenoeg alle droge Afromontane bossen in het noorden van Ethiopië zijn omgevormd tot open landbouwgebied. Er resteren slechts vele kleine geïsoleerde bosfragmenten rondom kerken ("kerkbossen"). Deze studie analyseert de structuur van deze kerkbossen, onderzoekt belangrijke knelpunten voor de regeneratie van houtige gewassen, en verkent de mogelijkheden en uitdagingen voor herstel. Dit promotieonderzoek behandelt de volgende hoofdvragen:

1. Hoe variëren de bosstructuur, soortensamenstelling en biodiversiteit tussen kerkbossen en wat zijn de belangrijkste factoren die dergelijke variaties veroorzaken? De factoren die dit onderzoek behandelt zijn hoogteligging, omvang van het bosgebied en de menselijke beïnvloeding.

2. Wat zijn de belangrijkste knelpunten in de regeneratie van houtige planten in kerkbossen? Deze knelpunten zijn onderzocht voor één tot zeven kerkbossen waarbij de nadruk ligt op de effecten van de zaadbank (in de bodem) en zaadpredatie op de vestiging van zaailingen, en op de invloed van begrazing, gradiënten in microklimaat en beheersmaatregelen op de vestiging, overleving en groei van zaailingen.

De structuur en soortensamenstelling van 28 bossen, gelegen op verschillende hoogten (1816 tot 3111 NAP) en variërend in omvang (van 1.6 tot 100 ha) is bepaald in relatie tot de hoogteligging, bosoppervlakte, begrazingsintensiteit en de intensiteit van houtkap. In de 28 kerkbossen die zijn bestudeerd zijn in totaal 168 houtige plantensoorten (100 boomsoorten, 51 struiken en 17 lianen) uit 69 verschillende families geïnventariseerd. Van deze soorten waren er 160 inheems en slechts acht exotisch (2 struiken en 6 bomen, afkomstig uit 6 families). Deze bossen herbergen veel soorten die slechts worden vertegenwoordigd door enkele individuen ('zeldzame soorten') en veel soorten die slechts in een enkel plot aanwezig waren ('unieke soorten'). Hierdoor verdient het instandhouden van kerkbossen prioriteit ten behoeve van natuurbehoud. De bossen verschilden sterk in soortenrijkdom (15 tot 78 soorten), stamoppervlak (4.8 tot 111.5m^2/ha) en stamtal (≥ 5 cm dbh: 267 tot 1553/ha; >1cm diameter: 619 tot 2421/ha en; zaailingen: 0 tot 5263/ha).

De belangrijkste factor die de soortensamenstelling van kerkbossen bepaalt is de hoogteligging. Onze resultaten laten zien dat verschillen in soortensamenstelling tussen deze bossen voor het belangrijkste deel bepaald worden door de hoogtegradiënt. De overeenkomst in soortensamenstelling was het grootst wanneer de hoogteverschillen tussen de bosgebieden minimaal waren. Het effect van geografische afstand op deze overeenkomst was gering. De enorme variatie in hoogteligging zorgt ervoor dat deze bossen de mogelijkheid bieden tot behoud van het grootste deel van de biodiversiteit in het gebied. Dit wordt bevestigd door het feit dat het aantal aangetroffen soorten in onze 28 kerkbossen (168) meer is dan het door South Gondar Zonal Department of Agriculture verwachte aantal soorten (125 houtige soorten). Anderzijds is de vegetatiestructuur van deze bossen vooral bepaald door de mens. De interactie tussen hoogteligging en menselijke beïnvloeding (met name door het weiden van vee) bleek verantwoordelijk te zijn voor de soortensamenstelling in de ondergroei. Niet alleen de soortensamenstelling, maar ook het aantal zaailingen was sterk beïnvloed door het weiden van vee. In een aantal bossen werden geen of slechts enkele zaailingen gevonden. Op lange termijn zal dit er waarschijnlijk toe leiden dat ook de soortensamenstelling in het kronendak deels het gevolg is van menselijke beïnvloeding. Het effect van de omvang van bosgebieden op de soortensamenstelling was niet significant. Dit impliceert dat, ondanks dat grote bosgebieden belangrijk zijn voor de regeneratie van houtige plantensoorten, kleine bosfragmenten in combinatie met een juist matrix beheer een goede aanvulling kunnen zijn voor het instandhouden van de biodiversiteit. Voor veel boomsoorten is de regeneratie in de meeste kerkbossen zeer beperkt. Inzicht in mogelijke oorzaken hiervoor werd verkregen door veldexperimenten.

Analyse van de zaadbank in zeven verschillende kerkbossen toonde aan dat zich in de bosbodem weliswaar grote hoeveelheden persistente zaden van kruidachtige soorten ophopen, maar dat slechts vijf (6%) van de 91 houtige plantensoorten die in de vegetatie van de zeven bossen zijn aangetroffen, ook aanwezig zijn in de zaadbank. De meeste boomsoorten accumuleren geen zaden in de bodem. Om te onderzoeken hoe zaden zich op lange termijn in de bosbodem gedragen bepaalden we de kiemkracht van de zaden van vijf boomsoorten op vier plekken in een bos na ze te hebben begraven voor 6, 12 of 18 maanden. Voor alle soorten nam de kiemkracht scherp af met de tijd dat de zaden onder de grond waren begraven, met uitzondering van *Juniperus*, waarvan de zaden na 18 maanden nog steeds 91% van de kiemkracht bezaten. Soorten verschilden significant in de kiemkracht na 18 maanden. De snelle afname van kiemkracht in de bosbodem geeft aan dat de zaden van de bestudeerde soorten niet persistent zijn. Het feit dat de meeste van de dominante boomsoorten geen zaden accumuleren die hun kiemkracht in de bodem behouden suggereert dat hun geslachtelijke voortplanting zal worden verhinderd door het verwijderen van volwassen individuen uit de vegetatie. Uit het predatie experiment op de zaden van zes boomsoorten in een van de kerkbossen bleek dat 92% van de zaden binnen 3-5 maanden ten prooi viel aan predatie. Een gebrek aan persistente zaden in de zaadbank, in combinatie met de hoge predatiedruk, bemoeilijkt daardoor de regeneratie in kerkbossen.

Deze studie toonde voor vier geselecteerde boomsoorten in twee verschillende kerkbossen aan dat de vestiging, overleving en groei van zaailingen ernstig bemoeilijkt wordt door vee begrazing. Nagenoeg geen van de uitgeplante zaden kiemden in proefvelden die opengesteld waren aan begrazing (de kieming was 4 en 5% in deze plots, vergeleken met 57.5 en 63.3% in proefvelden die waren afgescheiden van begrazing, data voor respectievelijk Dengolt en Gelawdios). De overleving van zaailingen in onbegraasde proefvelden was hoger (Dengolt = 65 en Gelawdios = 56%) en de groei van zaailingen was hier sneller terwijl in de begraasde proefvelden geen van de zaailingen overleefden. Dit geeft aan dat regulatie van begrazing door vee van groot belang is voor zowel de interne regeneratie van kerkbossen als het herstel van de gedegradeerde omgeving.

Deze studie onderzocht tevens hoe de regeneratie van bomen varieert langs een gradiënt van boskern tot bosrand en open gebied, en tussen gesloten en open kronendak in het bos. Zaailingen bleken zich het meest succesvol te vestigen in het bos en in het bijzonder onder gaten in het kronendak. Het blootstellingseffect tussen boskern en open gebied kwam niet overeen met het blootstellingseffect tussen gesloten en open kronendak binnenin het bos. Dit suggereert dat zaadkieming vooral beïnvloed wordt door de beschikbaarheid van water langs de gradiënt van boskern tot bosrand en open gebied. Ondanks dat zich binnenin het bos meer zaailingen vestigden was de groei van zaailingen sneller en overleefden meer zaailingen in de bosrand. Vooral randen binnen in het bosgebied bleken de regeneratie negatief te beïnvloeden terwijl zaailingen in de buitenrand van het bos juist hogere overlevingskansen hadden. Als gevolg van de interacties tussen licht en vocht zal de buitenste bosrand waarschijnlijk de optimale locatie vormen voor de overleving en groei van zaailingen. Mogelijkerwijs kunnen deze zaailingen de bos omringende en vaak gedegradeerd terreinen koloniseren en hun herstel bevorderen, mits het omringende land beschermd is tegen begrazing en andere landbouw activiteiten. Binnenin het bos werd de overleving en groei van zaailingen, zowel in droge als in natte perioden, gestimuleerde door licht dat door gaten in het kronendak viel. Mogelijk was in de droge periode de vochtbeschikbaarheid op open plekken in het bos hoger dan buiten het bos. Over het algemeen verschilde het kwantitatieve effect van de variatie in microklimaat met de soort.

Uit experimenten gericht op bosherstel door middel van actief beheer (zaaien, planten van zaailingen, onkruidbestrijding, verwijdering van strooisel, open maken van de bodem) bleek dat voor alle soorten de vestiging van zaailingen significant werd bevorderd door zaaien in combinatie met strooisel verwijdering. Dit duidt erop dat een gebrek aan zaden een van de knelpunten is voor de regeneratie van kerkbossen. De overlevingskansen en groei van zaailingen verbeterden niet door het verwijderen van onkruid, en had op een aantal soorten zelfs een negatief effect.

De toekomst van houtige planten en de kenmerkende vegetatie van droge Afromontane gebieden in Ethiopië hangt af van een effectief behoud en duurzaam gebruik van de resterende bosgebieden. Wanneer het niet mogelijk is alle resterende bosgebieden te behouden, dan is een selectie van kerkbossen variërend in hoogteligging essentieel om het merendeel van de soorten in het gebied te beschermen. Hiervoor is het van belang om levensvatbare populaties in het bos in stand te houden en bosgebieden met elkaar te verbinden. Het uitsluiten van vee en het verminderen van de houtkap zijn belangrijke voorwaarden om de regeneratie van kerkbossen mogelijk te maken. Daarnaast kan zaaien in combinatie met het verwijderen van strooisel en het open maken van de bodem het gebrek aan zaden compenseren. Het uitzaaien van plantenzaden dient gepaard te gaan met het actief reduceren van predatie en vraat (b.v. door het plaatsen van afrastering en het bieden van alternatieve voedselbronnen voor predatoren). Het planten van zaailingen is een ander alternatief om knelpunten met betrekking tot de zaadkieming, in het bijzonder vochttekort en zaadpredatie, te omzeilen, maar ook onkruidbestrijding kan nodig zijn. Het creëren van gaten in het kronendak binnenin het bos kan, afhankelijk van de soort, noodzakelijk zijn om de kieming van zaden en de ontwikkeling van zaailingen te bevorderen. Langs de gradiënt van boskern tot bosrand en open gebied blijkt zaaien vooral effectief te zijn in het microklimaat van de boskern, terwijl het planten van zaailingen resulteert in een betere groei en overleving van zaailingen in de buitenrand van het bos. Deze maatregelen stimuleren de regeneratie van boomsoorten. Het verbinden van bosrestanten door het aanleggen van vegetatiecorridors aansluitend op het bestaande natuurlijke terrein, of het verminderen van de onderlinge afstand door het creëren van bufferzones en plantages rondom de bossen, en de aanleg van groene landschapselementen zijn mogelijke beheersmaatregelen. Deze zullen de genenstroom instandhouden en uiteindelijk bijdragen aan het behoud van deze bossen en het herstel van het omringende landschap.

ACKNOWLEDGEMENTS

PhD study is a challenge. An art of facing and solving a challenge. In facing this challenge I was not all alone.

First and foremost I praise the Almighty God, who favors me to begin and to bring to an end this study.

The Archbishop of South Gondar Diocese Abune Elsa, the Manager Liqe-Tiguhan Abuhaye Fente with other staff were so kind to give me permission to go through the holy garden of the churches with their blessing and passion, thus there are no words to thank them. Farta, Laygaynt, Tachgaynt, Simada, Ebinat, Libo-kemikem, Fogera, Dera and Estie woreda church administration offices were very cooperative. The kind EOTC fathers, students and local community in each church who hosted and assisted me in identifying woody species, undertaking my field experiments are unforgettable and their wisdom is amazing.

Groneman BV and specially the director Ing. T.H. Sterck are gratefully thanked for their generosity to cover up most of my study expenses.

No words to thank my promoter Professor Frans Bongers, who went through each step of my work from draft proposal to the thesis work, from office to the field site and from logistic and financial matters to the scientific methodologies and skills. He is a person too sympathetic and inspiring to work with. Without his continuous support it would have not been possible to finish the thesis the way it did. I can not forget also the warm hospitality I got from his family in his home at different occassions during my stay in Netherlands. Dr Frank Sterck, co-promoter, is gratefully acknowledged for his dedication and time to go through all chapters repeatedly. His sharp and genuine comments have made me more analytical and precise in my work. Dr Demel Teketay, who always wished to see my success deserves my gratitude not only for his unreserved effort in this work as a co-promoter but also in all my earlier academic endeavors.

Forestry research Center (FRC) in Addis Abeba and the staff specially Ato Abeje Eshete, Ato Mengistie Kindu and Ato Bilehatu are thanked for providing me tree seeds for the experiments. Dr Tesfaye Bekele who contributed a lot during the initation of the project and helped me to get a vehicle for part of the field work deserves my gratitude.

I am in depted to Dr Lourens Poorter who read some of the chapters and gave his valuable comments and continuous inspiration. I also gratefully thank Dr Marielos Pena- Claros who not only assisted me in some of the survival analysis but also for her encouragement and inspiration. Dr Michiel van Breugel, my office mate with whom I shared the frustrations in finishing our study, helped me a lot in identifying and using some of the softwares to analyze my data and thus many thanks. Caspar Verwer deserves my appreciation who translated the summary in Ducth. I would like also to thank Ir. John Stuiver who devoted his time to plot GPS data of the church forests on map.

Thanks to many colleagues at the Forest Ecology and Management Group of Wageningen University for the support and friendship with whom I shared many good events: Frits Mohren, Toon Rijkers, Jan den Ouden, Ute Sass-Klaassen, Leo Goudzwaard, Ellen Wilderink, Neeltje van Hulten, Hans Jansen, Joke Jansen, Theo Ywema, Marco Dekker, Lars Markesteijn, Edwin Lebrija, Zhi-quan Cai, Emiru Brihane, Gabriel Muturi, Arnold van Gelder, Abeje Eshete, Paul Copini, Tefera Mengistu.

My gratitude should go to the Executive director of Organization for Rehabilitation and Development in Amhara (ORDA) Ato Wuletaw Haile Mariam and to all colleagues who allowed me to use all facilities in office and logistics during fieldwork. Their encouragement and after all their willingness to let me go through this study though I had unfinished working contract with them, is unforgettable. Ato Getu Hailu, Ibinat-Belesa project manager, and Ato Muluneh, South Gondar greening coordinator, are thanked for the support I got during some of the fieldworks. Ato Getachew Tamiru, Solomon Asres and Ato Fetene and Gebrhana Yalew are gratefully thanked in raising seedlings for the experiments. Estie woreda agricultural office staffs are also acknowledged for providing additional seedlings for the experiments.

Melake Hiwot Asnakew and his family hosted me so kindly while I was in the field (Gelawdios) for data collection. Their hospitality made my field work comfortable. My dedicated field assistants Nahu Asnakew and Assefa Lisanework are very much appreciated and thanked for recording data.

Ato GebreMeskel Fenta and his wife W/zo Bosena Melese with their family took care of my children while my wife and I were away for the thesis work. I would heartedly thank them for their unforgettable generosity and family-hood support. Had not been for them I would not have finished my study on time. My brothers

Ato Birhane Wassie and his wife W/zo Tsehay Yesmaw, Ato Ayalew Mengistu and his wife W/zo Bishat Ayalew were the ones who laid the foundation of my life from elementary school to university level and to whom I am very grateful.

Finally, my wife Sister Belainesh Melese who carried all of my family responsibility so that I could focus on my study is the corner stone in this work. Although in the mid of my study she went to university herself, she was very supportive and in many occasions she cared more for my work than hers. My little children Mahalet, Natinael and Yonas who were so patient and mature enough to be left alone while their mother and I were away for the study, are very special. The hardest thing in this study was to be away from them. Now, thanks to their endurance, it is done and we are going to be together again.

CURRICULUM VITAE

Alemayehu Wassie Eshete was born in South Gondar, Ethiopia, on the 6th of July 1971. He completed his high school in 1988 from Estie Mekane-Eyesus Comprehensive High School (South Gondar) and same year joined Alemaya University of Agriculture, Ethiopia (now renamed as Haromaya University). After four years of study he got his BSc degree in forestry with Distinction in 1992.

After graduating he joined the Ministry of Natural Resources Development & Environmental Protection as forestry expert and after one year promoted to be head of the district ('Woreda') office. His major task under this period was planning and managing the Forestry and Soil & Water Conservation programs and taking charge of all the administrative works of the office. In 1995 he joined ministry of Agriculture as leader of a team consisting of - Forest Expert, Agronomist, Irrigation Development Expert, Animal Husbandry & Fodder production Expert, Extension & Communication Expert in a district office.

From 1996 to 2000 he worked for an international NGO called Food for The Hungry International / Ethiopia (FHI/E) as Afforestation component Head in northern Ethiopia. As of February 2000 he joined Organization for Rehabilitation and Development in Amhara (ORDA) an NGO as forest development project coordinator.

In 2001 he joined Swedish University of Agricultural Science and got his MSc degree in farm forestry in 2002. His research work was assessing the role of Ethiopian Orthodox Tewahido churches in the conservation of forest Biodiversity in Northern Ethiopia. In this research he investigated Diversity and regeneration status of woody plants in church forests and socioeconomic aspects of these forests in South Gondar zone, Northern Ethiopia.

After finishing the MSc study he went back to ORDA and started working as senior forestry expert and later became head of planning and programming department of the organization. In May 2004, he became a PhD student at the forest Ecology and Forest Management group (FEM) of the center for Ecosystem studies of Wageningen University. In his PhD work he continued his MSc work on the ecology of remnant forests preserved by Ethiopian Churches which resulted this thesis.

Alemayehu Wassie is married and a father of three children.

PUBLICATIONS

PUBLISHED IN REFEREED JOURNALS

Alemayehu Wassie, Demel Teketay and N. Powell, 2005a. Church forests in North Gondar Administrative Zone, northern Ethiopia. Forests, Trees and Livelihoods 15: 349-374.

Alemayehu Wassie, Demel Teketay and N. Powell, 2005b. Church forests provide clues to restoring ecosystems in the degraded highlands of Northern Ethiopia. Ecological Restoration 23: 2.

Alemayehu Wassie and Demel Teketay, 2006. Soil seed banks in church forests of northern Ethiopia: Implications for the conservation of woody plants. Flora 201: 32–43.

Bongers, F., Alemayehu Wassie, F. J. Sterck, T. Bekele and D. Teketay, 2006. Ecological restoration and church forests in northern Ethiopia. Journal of the Drylands 1: 35-45.

SUBMITTED TO REFEREED JOURNALS

Alemayehu Wassie, Frank Sterck and Frans Bongers. Species and structural composition of church forests in a fragmented landscape of northern Ethiopia

Alemayehu Wassie, Frank Sterck, Demel Teketay and Frans Bongers. Tree regeneration in church forests of Ethiopia: effects of microsites, seed availability, litter and weed control.

PE&RC PhD Education Certificate

With the educational activities listed below the PhD candidate has complied with the educational requirements set by the C.T. de Wit Graduate School for Production Ecology and Resource Conservation (PE&RC) which comprises of a minimum total of 32 ECTS (= 22 weeks of activities)

The C.T. De Wit Graduate School PE&RC PRODUCTION ECOLOGY & RESOURCE CONSERVATION

Review of Literature (5.6 credits)

- Restoration ecology; forest regeneration; ecology of tropical forest remnants; the role of Ethiopian orthodox church in conserving forest resources in Ethiopia (2004)

Writing of Project Proposal (7 credits)

- Ethiopian church forests: opportunities and challenges restoration (2004)

Laboratory Training and Working Visits (4.3credits)

- Eco-physiology field experiment visit in Switzerland, Bern University (2004)

Post-Graduate Courses (4.8 credits)

- Basic and advanced statistics, PE&RC (2006/2007)
- Survival analysis, PE&RC (2007)
- Multivariate Analysis, PE&RC (2007)

Deficiency, Refresh, Brush-up and General courses (5.6 credits)

- Vegetation science and system ecology, Nature Conservation, Plant Ecology Group, WUR (2004)

Competence Strengthening / Skills Courses (2.3 credits)

- Leadership: on job training given by Organization for Development and Rehabilitation in Amhara (ORDA) in collaboration to Azusa Pacific University, USA, Home Institute (ORDA), Ethiopia (2005)
- Scientific publishing, PE&RC (2007)
- PhD competent assessment, PE&RC (2007)

Discussion Groups / Local Seminars and Other Scientific Meetings (4.9 credits)

- Workshop: “The global partnership on forest landscape restoration and the Netherlands” (2004)
- Workshop: ”Conservation issues in Dutch tropical research” (2004)

- Forest conservation and ecology discussion group (2004/2007)
- Ethiopian foresters association meeting and annual agricultural research reviews (2005/2006)

PE&RC Annual Meetings, Seminars and the PE&RC Weekend (1.2 credits)
- PE&RC introduction weekend (2006)
- PE&RC day: the scientific agenda: "who pulls the strings" (2007)

International Symposia, Workshops and Conferences (9 credits)
- SER 2004, the world conference on ecological restoration, Victoria BC Canada; presented a paper: "Ethiopian orthodox Tewahido church forests: their role in restoring forest ecosystems in the degraded highlands of northern Ethiopia" (2004)
- Earth in transition: how traditional ecological knowledge addresses global climate change part of world conference on ecological restoration in Zaragoza, Spain; presented a paper: "The contribution of Ethiopian church forests in combating ecological degradation an climate change effects in northern Ethiopia" (2005)
- International conference organized by The Association for Tropical Biology and Conservation (ATBC) in Morelia, Mexico; presented a paper: "Species and structural composition of church forest in a fragmented landscape of northern Ethiopia" (2007)

Supervision of MSc students: two students
- Church forest use and ecology: the impact of forest use on the woody vegetation in seven church forests in south Gondar
- Litter fall an seed rain in dry Afromontane forest relics in northern Ethiopia

Made in United States
Orlando, FL
05 November 2026

90200364R10128